Landside | Airside

Victor Marquez

Landside | Airside

Why Airports Are the Way They Are

palgrave
macmillan

Victor Marquez
Mexico City, Mexico

ISBN 978-981-13-3361-3 ISBN 978-981-13-3362-0 (eBook)
https://doi.org/10.1007/978-981-13-3362-0

Library of Congress Control Number: 2018965780

Cover image: Ho JJ
Cover design by Tom Howey

This Palgrave Macmillan imprint is published by the registered company
Springer Nature Singapore Pte Ltd.
The registered company address is: 152 Beach Road, #21-01/04 Gateway East,
Singapore 189721, Singapore

To my mother

and

In memory of Hans Meyer +

Preface: Current Definitions of the Landside and Airside

The Collins English Dictionary ("Landside, Airside," 2003) defines the terms "landside" and "airside" succinctly: "*Landside*" (noun) is the part of an airport farthest from the aircraft, the boundary of which is the security check, customs, passport control, and so forth; "*Airside*," on the contrary, is the part of an airport nearest the aircraft, the boundary of which is the security check, customs, passport control, and so on. Both definitions help to show that only an austere idea exists of what these terms mean, and that little research has been done to define them more accurately. In the general civil engineering and architecture lexicon, *airside* represents buildings and facilities on the side of the planes, and *landside* refers to the same but on the side of passengers, a definition similar to the one mentioned above. For the more specialized jargon of airport designers, engineers, and specialists, airports should be divided into two control sectors: the *landside*, referring to all areas allowing the free flow of passengers, visitors, and vehicles; the *airside*, areas restricted for the sole use of authorized personnel, aircrafts, and service vehicles.[1] However, in terms of airport planning, security, and regulatory codes, the landside–airside boundary represents the different frontiers between the "*sterile*" and the "*non-sterile*" zones, a through area where passengers move along "*filters*" in order to be "*cleared*" and "*segregated*."

Mexico City, Mexico Victor Marquez

Reference

Evans, G. (2006). *Airport engineering: New Denver Airport, Colorado*. New York, NY: McGraw-Hill Encyclopedia of Science and Technology.

[1]Facilities at airports are generally described as either airside, which commences at the secured boundary between terminal and apron and extends to the runway and to facilities beyond, such as navigational or remote air-traffic-control emplacements; or landside, which includes the terminal, cargo-processing, and land-vehicle approach facilities (Evans, 2006).

Acknowledgements

Any intent to produce serious writing requires the inspiration, intelligence, and long dedication of its author; however, authorship requires meeting at least two conditions: first, the humility to acknowledge that we only interpret from the experience and lives of others and that it is only because others have produced knowledge before us that we can see further; second, that within the darkness of that accumulated sea of thoughts, letters, and images, it is only through the guidance of others that we may find the path toward what we seek. Hence, what we put on paper, as individuals, will always remain in time as a collective effort. This book is precisely that, a collection of ideas of many talented people.

Nonetheless, I should acknowledge many friends and colleagues in particular. I first thank Peter McCleary—my intellectual guide in the second half of my life and a teacher who introduced me to the deepness of the philosophy of technology. Without that first impulse, what came next was near to unthinkable.

Very specially, for his patience and intellectual generosity I shall thank Dr. Ron Kline, who became my mentor after we crossed words in 2005. This work would not have been possible without the long hours Ron devoted to marking up the manuscript and the several extra hours he gave advising me in his office, a café, or anywhere else we met in the world. Furthermore, I owe him because through the years he has showed me a new way of thinking: Relativism. Likewise, I show my gratitude to Dr. Trevor Pinch, Dr. Mike Lynch, and Dr. Phoebe Sengers from Cornell University for their invaluable contributions and advisory to build this book.

Much of the core material that set out the basis for my argument came directly from those who, with their talent and determination, established the profession of airport planning. I cannot describe how thankful I am to the legendary Marge Brink Coridan, Gene Lewis, Alistair Sherret, and Bill Dunlay for answering my questions and, along the way, passing on to me their vibrant enthusiasm for airports. I especially thank my friend Alceste Venturini for opening the doors of Jacobs Consultancy and its literature, truly an invaluable gesture. But the further back I went in history, oral histories were silenced by the inexorable pass of time. Thus, I needed the valuable assistance of those who order and classify knowledge. I appreciate the help I received from Douglas Di Carlo at La Guardia and Wagner Archives in Long Island City, Cynthia Ostroff at the Manuscripts and Archives at Yale University Library, Rodney Kite-Powel at the Tampa Bay History Center, Clarice Reardon at the Hillsborough County Library, Christine Peers and Jenifer Peals at the Tampa Historical Society, Dr. Robert Kerstein at the University of Tampa, Dr. Gary Mormino at USF St. Petersburg, and the staff at the National Archives II in Maryland, the New York Public Library, and the Avery Architectural and Fine Arts Library at Columbia University.

In this work, I often echo the voices and work of others, and they deserve much of the credit for this book. I kindly thank those authors who lavishly shared thoughts about this work or discussed with me topics related to airport history such as Dr. Deborah Douglas—an inspirational model for this book, Dr. Marc Dierikx, Dr. Peter Lyth, Dr. Joe Corn, Dr. Nathalie Roseau, and Dr. Gordon Pririe. I also thank my colleagues Angeles Miranda, Graciela Torre, Jean-Philippe Percheron, Emilio Gil, and Jorge Sarralle for those long discussions in the practice of airport planning and design: We were in fact, from the professional trenches, chasing the meaning of what a landside–airside boundary might be. We still are.

Any book is also a collection of words, which must be carefully placed and hierarchized in order to transmit ideas and thoughts. Thus, I owe much of the structure and readability, coherence and style of this text to some generous friends, teachers, and editors who directly or indirectly have contributed to this work. I thank Dr. Keith Hjortshoj because he showed me the power of poetry, narrative, and style. I also deeply thank my lifelong friend Sabrina Spannagel at the University of Washington for her advice on the first chapters of this piece. Perhaps the biggest practical input came from my style editor, Kristen Ebert-Wagner, who shoulder to shoulder discussed with me each idea and paragraph, until it reached its final shape. I have been privileged to learn from all of them. Finally, I sincerely appreciate the patience and support of Joshua Pitt and everyone at Palgrave Macmillan in London, without them this work would never have reached the print.

Contents

1 Introduction 1
References 19

2 "The Romantic Borderline": From Fences to the Skywalk—Landside–Airside Space in an Early American Airport; New York LaGuardia Terminal 1933–1939 21
1 Introduction 21
2 Pilots Are Gods Descending from Heaven 23
3 The Fence 27
4 The Romantic Borderline 31
5 Reinventing an Airport in 1937: "Country Club" or "Rail Station"? 39
6 The "Soft Transition" 48
7 The End of an Airport's Epoch 62
References 66

3 "Reinventing the Airport?": *Annex 14*, Dulles Airport's "Mobile Lounge" and Other Jet-Age Paradigms, 1946–1962 71
1 Post-Second World War Airports 73
2 Formalizing Airport Planning 80
3 The Language of Planning 85
4 How Did Airports Gain "Technological Momentum"? 91
5 Toward the Reinvention of the Airport 99
6 The Birth of the Mobile Lounge 102
7 Materializing a New Landside–Airside Boundary 113
References 121

4 "The Landside Airside Concept": Breaking to Reconnect—The "People Mover" at Tampa International Airport, 1962–1971 125
1 Framing Tampa 126
2 The Airport and Its Actors 128
3 Problem Posing 144
4 Problem Construct 157
5 Problem Solving 165
6 After the "Landside Airside Concept" 168
References 172

5 "Are Landside–Airside Boundaries Cultural Mirrors?": Reinventions, Innovations and Society 175
1 On Airport Reinventions 177
2 On Airport Innovations 182
3 The Cultural Mirror 191
References 201

6 "The Liquid Airport": Security, Permeability, and Containment in Airports 203
1 Discursive Aspects: The Paradigm of Security, Rethinking What Airports Are 205
2 Normative Aspects: Coding a *New Air World Order* 213
3 Constructing an Object of Knowledge 216

4 Transitory Aspects: The Limbo 224
5 Intangible Aspects: The Immaterial Frontier 229
6 Boundaries and Beyond 238
References 240

7 Conclusion 245
References 251

Postscript 253

Bibliography 257

Index 277

About the Author

Dr. Victor Marquez is a writer, researcher, and designer interested in urban theory, technology, mobility, and culture. He earned a Ph.D. in science and technology studies from Cornell University in New York and also holds a Master of Arts in sociology and history of technology from the same institution, a Master of Architecture from the University of Pennsylvania, and a Bachelor of Architecture from the National University of Mexico UNAM. As a Fulbright scholar, he has conducted research on airports and the culture of technology for more than ten years. He has also written and lectured on mobility, systems, and philosophy of the futures of technology for *The Futurist*, Discovery Channel and as a Ted Talk speaker. In 2016, he was one of the two candidates for the National Prize of Exact Sciences from the US-Mexico Commission. As an architect and planner, he has contributed in more than twelve airports including L. B. Pearson in Toronto with MSA, and his own designs for Monterrey International in Mexico, and ten more across the region. He is a member of the Urban Design Council and the Parks Commission of Mexico City. He is also a member of the Society for the History of Technology (SHOT), the International Association for the History of Transport Traffic and Mobility (T2M), the Society for the Social Studies of Science (4S), the Swiss Association for the Studies of Science, Technology and Society (SASTS), the Cornell Club in New York City, and the Wharton Club in Mexico. He has taught in Australia, the USA, and Mexico and given more than fifty lectures and papers in five continents. He currently lives in Mexico City.

Acronyms

ADP	Aéroports de Paris
AOPA	Association of Pilots & Aircraft
CAA	Civil Aeronautics Authority
CWA	Civil Works Administration
DHS	Department of Homeland Security
FAA	Federal Aviation Authority
IATA	International Air Transport Association
ICAO	International Civil Aviation Organization
L&B	Landrum & Brown
LF&A	Leigh Fisher & Associates
LOS	Levels of Service
PCP	Passenger Collection Point
PICAO	Provisional International Civil Aviation Organization
ROW	Rights-of-Way
S&TS	Science & Technology Studies
SASTS	Studies of Science, Technology and Society
SHOT	Society for the History of Technology
SOM	Skidmore, Owings and Merrill
WPA	Work Projects Administration

List of Figures

Chapter 2

Fig. 1 Two-level split at LaGuardia Airport, separating for the first time departures and arrivals. A break though innovation that would make a huge impact in airports worldwide. *La Guardia Airport Passengers* [digital] (1944) (Reproduced with permission of Underwood Archives/via Getty Images) 22

Fig. 2 A key example of the early landside–airside boundary. Spectators are separated and contained in fenced areas. Leslie R. Jones, Photographer (ca. 1930) (Courtesy of the Boston Public Library, Leslie Jones Collection) 29

Fig. 3 A view from the Skywalk platform looking toward the apron. Notice the conceptual split between the platform and the parking. Wurts Bros. New York, NY (1944) (Image provided by the Museum of the City of New York) 55

Fig. 4 LaGuardia pioneered in separating the different flows. This view shows the courtyard formed between the Terminal and the Skywalk promenade. Notice some ground support vehicles at the forefront, passengers circulating underneath,

and visitors walking on top of the structure (Photography by The Port of New York Authority (1955). Donated to the Museum of the City of New York) 57
Fig. 5 This plate of LaGuardia Airport pictures the thousands gathered at the opening day. *Dedication of New York Municipal Airport* (1939) (Reproduced with permission of *NY Daily News*/via Getty Images) 65

Chapter 3
Fig. 1 Artist rendering of Dulles Terminal by Eero Saarinen. This is one of the few modern buildings ever depicted on a US stamp. Solodov, Aleksei (1982). *A stamp printed in the US shows Dulles Terminal* [digital] (Reproduced with permission of Shutterstock) 72
Fig. 2 Readers should consider the technical difficulties that different aircraft heights pose to engineers. The so-called gangplanks needed to mechanically adjust the levels and at the same time keeping the slopes comfortable for passengers. Notice the invention of the "telescoped" system. Fenn, Al (1958). *A scene from O'Hare Field, at the United Airlines landing strip, as passengers walk down the interior of the gangplank which will mount between plane & building* [digital] (Reproduced with permission The Life Picture Collection/Via Getty Images) 97
Fig. 3 A new kind of apron reinvented for the future. Korab, Balthazar (1963). *Aerial view showing mobile lounges* [digital] (Unrestricted use. Balthazar Korab Archive at the Library of Congress) 109
Fig. 4 Commercial postcard showing how the passenger's upload to the aircraft now takes place in the middle of the apron. Korab, Balthazar (1958). *Mobile Lounges* [digital] (Unrestricted use. Balthazar Korab Archive at the Library of Congress) 111

Chapter 4
Fig. 1 A view of the centralized "landside" building and its "airside" satellites. Notice the interconnecting piers bridging the hard frontier between apron and road. *Aerial view of Tampa International Airport* [digital] (1972). Public domain (Provided for reproduction by the Florida photographic collection at the Florida State Library) 129

Fig. 2 People movers remain in operation after thirty-five years of its construction. Howes, James G. (2009). *People mover connecting Terminal A to the central terminal* [digital] (Reproduced with permission of James G. Howes) 142
Fig. 3 A view of the original complex scale-model by Reynolds, Smith and Hills in 1970. *Physical model of Tampa International Airport* [digital] (Image courtesy of the VMA Archive) 145
Fig. 4 View showing the mechanized connector between the landside and airside buildings. *Bird's-Eye View of Tampa International Airport* [digital] (1973). Public Domain (Provided for reproduction by the Florida Photographic Collection at the Florida State Library) 149
Fig. 5 Gene Lewis' sketch illustrating how Tampa's concept of radial satellites—shown on the left, evolved into Atlanta's idea of a sequence of modular terminals. The author makes emphasis on the mechanized connections in both cases. Eugene Lewis (2008). "From Tampa to Atlanta" Ink drawing on paper (Courtesy of VMA Archives) 155
Fig. 6 These diagrams show the different possible layout arrangements for a landside–airside separation. Leigh Fisher and Associates (1963). "Basic Concept Diagram, Landside–Airside Concept" by Brink Coridian, M. (June, 2006) in Leigh Fisher Associates: History and Lore, as see through the eyes of Marjorie Brink Coridian (Courtesy of Marge Brink Coridian) 158

Chapter 5
Fig. 1 An artist's representation of airport morphology, showing the diversity of forms within this building typology. The world's 30 busiest international airports are presented as terra cotta bas-reliefs, hand-built to the same scale and all oriented to True North. Abrams, Janet [Pieces in terra cotta. Installation: 10 ft × 16 ft] *In the Unlikely Event* (2013) (Reproduced with permission of Janet Abrams [www.janetabrams.com]) 176
Fig. 2 Rehearsing a traditional serenade, typical in Mexican welcomes, even after short trips. Scherschel, Frank (1958) *Central airport where musicians entertain travelers* [digital] (Reproduced with permission of The Life Picture Collection/via Getty images) 199

Fig. 3 Are airports turning into the new museums? Reinhard, Tiburzy (2017). *The sculpture "The Story of Fog Woman and Raven" by artist Dempsey Bob is permanently exhibited in the terminal of the Vancouver International Airport* [digital] (Reproduced with permission of Shutterstock) 201

Chapter 6

Fig. 1 Cutting edge technologies, like metal detectors, are used for the first time in London's main airport. Observe the gendered connotations of the device. *New Anti-hijacking* [digital] (1970) (Reproduced with permission of Popperphoto/Getty Images) 204

Fig. 2 The dance of the "docile bodies." Dobilas, Marius [digital]. *Security and passport control at Antalya Airport* (2017) (Reproduced with permission of Popperphoto/Getty Images) 207

Fig. 3 The civil airport is now turned into a militarized facility. Frame Stock Photographs [digital]. *Airport security apprehending a suspect* (2015) (Reproduced with permission of Shuttersock Inc.) 222

Fig. 4 Limbo is best represented by reality, how someone end up living nowhere? Fougere, Eric [digital]. *Sir Alfred Mehran* (2004) (Reproduced with permission of Corbis Entertainment/via Getty Images) 224

List of Tables

Chapter 3
Table 1 Actors who intervened in the decision process for Dulles, by main interest and specific role during negotiations 115

Chapter 4
Table 1 How Tampa International Airport Meets Bijker's criteria and how it may be analyzed as a technological frame 128

Chapter 6
Table 1 The five epistemic features that may shape a technological viewport 217

List of Boxes

Chapter 4

Box 1	Advertisement for Hillsborough County Aviation Authority, Published in *Tampa Tribune*, 1971	142
Box 2	Advertisement for Eastern Airlines, Published in *Tampa Tribune*, 1971	142
Box 3	Glossary from *Recommended Planning Criteria for New Passenger Terminal Facilities at Tampa International Airport*	147

1 Introduction

Before I move on to the book's contents, I have to explain the reader the format I have chosen for the book. Although I have included some key illustrations, printing on paper has several limitations compared to the wonderful possibilities of electronic media. So, I invite the reader to visit the webpage www.landisdeairside.com where you will be able to see both the illustrations that are reproduced in full splendor, along other fascinating historical photographs, documents, and drawings. It is important to say that the Web site is a separate entity, fully non-profit and built only with the purpose of amplifying the discussion and analysis of the topics related to airport history. It has been said that one image speaks a thousand words; in this book, this may be true, but I will do my best to gear you up with a special pair of glasses that eventually, will allow you to see beyond the images. Precisions made, we move on to the difficult task to understand, why airports are the way they are.

The history of the airport seems to be rather different from that of its relative: the train station. While both aircraft and rail served as mass means of transport during most of the twentieth century, it is remarkable to see how similar train stations are around the world and to witness how most of them work really well. They are practical, easy

V. Marquez, *Landside | Airside*, https://doi.org/10.1007/978-981-13-3362-0_1

to understand, and efficient in expediting multiple convoys of trains, with precision down to the second. Most of these facilities share a similar configuration, which has been proved right for a long time, creating a sort of basic design pattern for rail transportation planners. One may infer that in socio-technical terms, train stations rapidly reached a "design closure" (or when experts agree that the basic problem of planning has been solved and thus left behind). In contrast, since their origins, the development of airports seems to be full of uncertainties. Just a glimpse of the last eighty years of airport design showcases a surprisingly wide catalog that does not seem to have two similar boundaries. Although hundreds of airports share a similar layout, today there is hardly such a thing as a typical airport; instead, each one of them is so puzzle-like, so aesthetically diverse, that problem-solving skills are demanded of passengers and visitors. Figuring out where to go, what to do, and how do it on time is the norm of contemporary air of travel. Clearly, airports have become much more complicated than rail stations, and this is still fairly inexplicable. And while this question has remained unresolved for decades, the so-called level of service that rules both the aviation and airport industries has fluctuated decade by decade, year by year, just like the stock market, ranging from the true excitement of many to the more exaggerated air-rage syndromes of some. Rail stations have successfully sorted out congestion and delays, whereas airports have historically undergone serious jam crises, and we as passengers have paid the price of these inefficiencies.

I personally became concerned with this matter, or at least aware of it, when I started working as an airport designer in 1998. Since then I have become increasingly involved in both airport and transportation planning projects and have developed an interest in their historical development. But it was not until my second year in Cornell's super demanding Ph.D. in Science and Technology Studies (S&TS) program when I started looking how technologies are perceived differently according to context and the social network that creates them. Because this book is thoroughly permeated by S&TS ideas, I think is relevant to say that I interpret S&TS as a new, interdisciplinary, and systemic field of study that analyzes the social, political, and cultural factors that shape science and technology, and how society is not necessarily driven by technological advancements. I must confess that it took me no less than two years in order to fully grasp this concept, and it was then was then when I realized that in essence almost no two airports looked the same.

Further inquiry suggested that most airports needed to be tailored-designed, thus making difficult the creation of standard models. This condition also seemed to contradict the very stable technological change in aircraft—including the change from propellers to jet engines after the Second World War. Indeed, one may argue that aircraft innovation was mainly invisible to users and related less to aircraft shapes than to their technical specifications, safety systems, and speed. Throughout history, airports have successfully increased and adapted the capacities of the airside in order to meet dimensional changes that are the product of technological change. But it was not until I became more aware of the mechanisms that allow a social construction of technology and of the ways that social historians put them in context that the invisible started to become not just visible but evident. There seems to be a single obstacle to airport standardization or patterned-continuity: The serious conflict that arises in defining the frontier between the landside and the airside.

In 2006, while I was in charge of designing a new international terminal for an airport, I noted that most of the planning meetings stalled when we were trying to define the overlapping of zones between the landside and the airside. In those meetings, everyone had an opinion about how to draw the line between the two, so consensus was hardly reached. At the time, I became more attentive to these issues and tried—unsuccessfully—to find references in the available literature on airports. But every path became a dead end, sadly, there was scant record of the frontier as an object of scholarship. In a search to amplify the concept, I borrowed some key S&TS concepts and methods for my inquiry. Perhaps it was Star and Griesemer's (1989) concept of "boundary objects," a classic within the field that became the closest reference on my way to figuring out what a landside–airside boundary could be. Although the concept was not necessarily applicable in the same way, I found some interesting common threads; in practice, just as "boundary objects" function—working as a point of linkage between diverse specialists—landside–airside boundaries are right at the crossroads of not just specialists but institutions, the state, the prevailing ideology, and the larger discourse of society. However, they lack integrity because the concept of the landside–airside boundary is not necessarily used indistinctly on a regular basis as a means of expert communication. Thus, I learned that although they are always present in every negotiation, these boundaries are slippery entities, potentially hard to trace, identify, or even make visible. However, as I suggest through this work, they are often tangible objects.

In hindsight, it all started making sense. I gathered my own recollections and notes starting from the days I worked as a project architect with Moshe Safdie and Associates in Boston; some colleagues and I were in charge of developing the expansion of Lester B. Pearson's Toronto International Airport. Indeed, most of the time we worked—rather than with the architecture or with the endless technicalities—using simple, colored diagrams to represent jurisdictional zones. Some of these zones were symbolically owned by different entities or showed the status of passengers and employees, others marked the functional separation between retail and seating areas, and yet others represented flow patterns, restrictions, and codes. We would pin up and overlap those diagrams and then hold meetings, often packed with consultants and experts; we always upended trying to sort out the near-to-impossible task of pleasing everyone with an integrated diagram. I clearly recall that the need for a double segregation flow—unique to Canada, according to which US passengers are split from Canadians and again split from all other nationalities—seemed literally insurmountable. We spent months zoning and eventually designing the three-dimensional boundaries that will end up "segregating" those passengers. We devoted most of our energy to defining what I now call the landside–airside boundary, but back then, we were not even aware of it.

By Fall 2008, I decided to move on to formally researching this topic. After several months of reading, I could confirm that landside–airside boundaries were not discussed in most airport literature, so I felt compelled to investigate whether other groups of colleague planners, designers, and experts had had the same experience and trouble in defining them as physical entities. My first approach was historical, but instead of following a chronological order, I forced myself to randomly select airports and, reading the subtext of both text and image, try to reveal what was behind the surface, pushing hard to identify in each case where the landside–airside was. As some cases led to others, everything pointed to a few bundles of airports that were truly innovative. What happened next became the core of this work: a fascinating journey into the past, seen through the eyes of the boundary, where airports reveal a new side, full of the traces of those who shaped them, and why. But what if—as I try to show in this book—the landside–airside boundary *is* the single most important feature that shapes an airport? Can we challenge the way we have written the history of airports and the form in which we understand them today and perhaps tomorrow? This question is precisely what this investigation tries to answer.

For the purpose of this inquiry, the landside–airside boundary will work as a filter or a lens that may help us build a different notion of what airports are. At the beginning of my search, it seemed clear that the first challenge was to come up with a comprehensible definition; but in this case, just establishing the parameters of what to look for would build up the foundations for a research hypothesis. I was disappointed to find that most available definitions are reduced to flat descriptions of the physical condition of a landside area or an airside area. Not even in the milestones of planning such as Froesch and Prokosch's *Airport Planning* (1946) are the terms mentioned. Therefore, I have opted to weave together from a variety of technical threads, ranging from my own interactions with specialists in the practice of planning and designing airports to books and other texts (mainly old ones), some interesting definitions contained in airport codes, and so forth.

First, I would suggest that landside–airside boundaries are sociotechnical objects that are often materialized physically as a line or a fence, or as an artifact such as a net, chain, or queue divider; they sometimes appear as a wall, as signage, or as a glass panel. They also shape as scanners, or ID documents, or metal detectors. These boundaries may also take non-material shapes such as routines, procedures, jurisdictional limits, the containment or "segregation" of passengers, or control demarcations of higher entities such as agencies, customs, or an assortment of police divisions. Boundaries may be even embodied in psychological emotions such as the public perception of fear, scrutiny, or the invasion of privacy. But they are all socio-technical objects that have as their ultimate purpose the enforcement of the separation of entities, mainly between the realm of aircraft and the realm of persons. In any of their variations—or multiple disguises, to say it a bit more theatrically—landside–airside boundaries also mark the frontier between the local and the regional, or the national and the international. Therefore, these boundaries have multiple functions of mediation and containment, and in consequence, they usually impose some sort of mechanism of control.

At the outset of this study, I tried to hypothesize the representation of an *objectual* landside–airside boundary—mainly because in practical terms I needed to make it visible and classifiable. However, as the research moved forward I learned what makes the boundary so hard to trace: on one side, its immutability as a *socio-technical* interface, and on the other, its constant and enigmatic change of skin.

Why is the boundary relevant and to whom? I suspect there are two ways of answering this question. On the one hand, there is an imperative need to better understand how to plan and expand airports, reducing the huge economic risk that is implicit in such monumental investments (a new airport costs between 6 and 14 billion dollars, Rodrigue, Comtois, & Slack, 2009, p. 2). Thus, the need to learn from the past seems obvious, and I believe we have not done so very successfully. Therefore, we, as scholars, need to expand not only the literature of airports just for the general public but also the specialized literature that is used by airport planners and designers, engineers, sociologists, and psychologists of transport, politicians and airport managers, and others. Historians of technology should try to unthread airport history and make the invisible visible, and thus part of a wider body of knowledge. On the other hand, let's imagine for a second how a historical evolution of airports might have occurred. Had airports moved along a technical progression—through years of refining layouts and learning by using (Rosenberg, 1982)—probably sooner than later planners would have been able to nail down the *right airport*, or at least an ideal way to expand airports? But this has not been the case, and the history of airports has been a collection of beautiful, even whimsical buildings—or, paraphrasing Perlman (2004), alongside museums, the highest trophy for star architects. This may explain why books of airport history are less text and more illustration: There seems to be little to say about airports beyond architectural critique. Besides the natural expert audience for this text, I think the identification of the landside–airside boundary may be relevant to the scholarship of S&TS, because it is only through understanding how these artifacts and systems are socially constructed that we can conceive an unconventional history of airports.

Among the various obstacles to the identification of the landside–airside boundary as a key element in airport history, outstanding is that consulting firms, particularly airport planning companies, have jealously vaulted most evidence and accounts of any negotiation related to its formation.[1] This would partially explain why there is virtually no public access to these sorts of records, and why the task of unveiling these internal processes is so hermetic to outsiders. Thus, it is not until researchers try to become fully immersed in a kind of participatory research (or until the researcher has somehow managed to join the negotiation processes) that these "black boxes" may be finally opened up for inquiry. However, the researcher must be cautious about discussing this knowledge, which companies often seek to protect in order to charge high consulting fees. But even supposing that a researcher has sorted out this conflict, it would be necessary to breach confidentiality agreements. This tortuous chain of steps has made this task an almost impossible one.

In regard to this study, though, several things lined up in my favor. From the time I began thinking about landside–airside boundaries at Cornell until the final days of typing this text, I have had the chance to participate professionally in both new airport projects and airport expansions. Within that context, I was fortunate to design, under my own office, the new Monterrey International Terminal B in Mexico. We have planned, designed, and overseen the project's full-time line—which spanned from drawing the first concept sketch to helping to put it into operation. This became an invaluable opportunity, because

[1]Planning companies have treasured this information basically because it is critical for developing any sort of airport master planning involving terminals. Thus, it is one of the reasons that justify the high fees of these experts.

I could act as a participant in being two different roles (designer and researcher) throughout the full negotiation process for, implementation of, and early use of the landside–airside boundary.

Although airports as we know them today were "invented" around 1920, there were no archetypes for the new architectural genre. During those years a myriad of architects and urban designers—including high-caliber figures such as Le Corbusier, Tony Garnier, Frank Lloyd Wright, Antonio Sant'Elia, and Erich Mendelsohn—all of whom were excited by the new possibilities of flying, launched a cascade of science fiction proposals that tried to envision airports of the future. However, all of these hypotheses were eventually proved wrong and became no more than wonderful visions of dystopia. In a different arena, the rising market of aviation pushed less fancy names into the spotlight and became finally responsible for bringing about down-to-earth buildings such as Tempelhof, Croyden, Le Bourget, Liverpool, and Dublin. Consequently—with the notable exception of Berlin's *flughaven*—these air terminals were far from grandiose and spectacular. Since their outset, airports faced problems that were somehow unexpected. Three key conditions separated the train's speedily service from its relative, the airplane: first, the uncertainties and complications of human maneuverability (extremely limited in twentieth-century train systems); second, the slow time for embarking, preparing, and disembarking of each aircraft (also called turnaround time); and, last but not least, the dangers involved in allowing people proximity to propellers.

The most spectacular airport seen before the Second World War was certainly Tempelhof Airport. Historian H. J. Braun (1995) also describes it as a technical and cultural feat: "from a symbol of efficiency in air transport in the rationalization period of German industry in the latter half of the 1920s, it changed to a symbol of Nazi grandeur in the 1930s" (p. 45). Its fame has prevailed to the present, making Tempelhof a silent witness to German history. The Berlinese airport was unequaled in many ways. In plan, the complex was organized along the edges of a gigantic elliptical plot, quite uncharacteristic of today's standards. The layout divided the airfield with a hangar-like, semicircular overhanging roof, where aircraft uploaded and downloaded passengers as on a loading platform, which bled into the apron. Within this monumental space, stair ramps divided the flow of passengers from service personnel. At the landside, sets of endless severe Art Deco buildings held offices and vestibules articulated by Albert Speer–like colonnades,[2] also surrounded by immense parking bays and road systems. The layout of the airport was so characteristic that it became a cultural icon (Zukowsky, 1996, p. 28; see photograph on p. 59). During the early 1930s, most terminals were simple, symmetrical buildings housing centralized functions of ticketing, waiting, and services. The immediate consequence of growth was a need for larger, longer aprons in order to park more and more aircraft. Between 1934 and 1937 George Labro, who was in charge of expanding Paris Le Bourget, did the most obvious and foresaw one long building, centrally divided by a core, expandable at both ends. However, the structure ended up being a quarter of a mile long. This idea would become a paradigm and thus the more common way to expand existent airports worldwide, even before the Second World War. This should have been the end of this story, but two facts changed the historical trajectory of this neat technological scheme: First, passengers were unhappy with the discomfort produced by walking such long distances, and second, the linear growth of airports was limited to their total land property (especially given that these facilities were located relatively near to city centers). In consequence most early airports became rapidly obsolete. I argue, then, that it is right when airports "needed to be reconceived" that we might be able to understand their lack of technological stability (we may compare the genre of the airport with cinema halls, and see how much theaters' layouts have remained stable through the same last century).

With the previous idea in mind, and after reviewing the international history of airports from the late 1930s until the 1990s, I found that only three airports have been rethought from the ground up: the 1939 NY–LaGuardia Airport, the 1961 Washington–Dulles Airport, and the 1971 Tampa International Airport, all of which happen to be in the USA. I also found that all three share a common thread: A higher entity specifically ordered a full reinvention of the airport. Interestingly, these

[2]An evocation of the *Zeppelintribune* by Albert Speer.

three *reinventions* reached only a rhetorical closure—meaning that the problem of the airport was never totally solved. This historical "irresolution" brought the landside–airside conflict into the light. Hence, these three cases became rich queries for understanding how a reinvention of the airport consists in a full reformulation of the landside–airside boundary.

Thus, in each chapter of this book, I have carefully analyzed, unthreaded, and unfolded all three cases in order to make visible actors, their interests, and the negotiations that took place in shaping their own frontiers. I have divided this research into four chapters. In Chapter 2, I present early examples of the landside–airside boundary and comment on its origins and representations. I also discuss the emerging complexities that turned airfields into airports and airports into systems. Here, I argue that along the transition from the aerodrome's leisure character to the more utilitarian airport, a deep gap was drawn between Corn's characterization of people's "airmindedness" and the rise of a new technical culture that arrived nested within the official discourse of modernity. In those years and from the point of view of users, the *romantic borderline*—or the earliest representation of the landside–airside boundary—was in danger. As aircraft grew more sophisticated, airports became a technology of their own; they turned into political artifacts and powerful mechanisms of publicity and propaganda. To cover its need for a new airport, in 1939 New York City officials commissioned a new kind of airport that could surpass its rival, Newark Airport, in New Jersey. Not surprisingly, the design incorporated a never-seen-before landside–airside boundary scheme. In this chapter, the NY Municipal Airport, later known as LaGuardia, is analyzed in depth, with a special emphasis on the larger discourse and its conflict of interests. In addition, this section also discusses the details of the creation of the "Skywalk," or the birth of the first techno-political landside–airside boundary. In analyzing the phases of funding, planning, design, and construction, I discuss the technical and political rhetoric of the time, which evidences a complex negotiation between the interests of users, machines, and the state. I suggest that the "Skywalk" became a *soft frontier* between two collapsing ideologies (soft as a synonym of mediated or avoiding a drastic rupture).

Chapter 3 introduces the airport culture of the post-Second World War period as a framework for understanding the growing demands of aircraft innovation (jet propulsion) and the implications of the air travel boom of commercial aviation during the 1950s and 1960s. As the number of airplanes grew, so did the number of passengers, in the same proportion, pushing the terminals to lengthen endlessly as long docks. Passenger discomfort became a constant; thus, saturation and delay problems took over the airport. Designers and experts responded with all sorts of mechanisms such as the 1952 "Loadair System" by Whiting Co., the 1954 "Air dock" prototype by United Airways and the installation of the first "Jetway" by JBT Co. in 1959. Meanwhile, politicians and investors were more concerned with the public image of the airport as an icon of progress. Terminals were identified as city gates, rapidly turning into the new urban cathedrals our time. The iconic TWA terminal in New York was inaugurated and instantly praised as an architectural masterpiece. A few years later the building's sound public success as an urban landmark, placed Eero Saarinen, at the forefront of the new Washington airport's public search for an architect. This time, the recently created Federal Aviation Authority (FAA) pushed an initiative to develop new ways to deal with a severely congested landside–airside boundary, and Saarinen was chosen for this task. The Finnish architect faced the question of how to reduce delay, and his answer was to rethink the airport as a system. Chapter 3 in particular emphasizes the experimental planning behind the new Washington–Dulles Airport in the late 1950s. In this chapter, I also recount how the era's faith in technology as a vehicle for progress was the main force behind the invention of the "Mobile Lounge" and how this artifact was seen as the ultimate solution for the ideal airport. Using primary and secondary sources, this study goes deep in search of the causes of and ideas behind a radical reformulation of the landside–airside boundary. Finally, I try to answer the question of why the FAA (US's Federal Aviation Authority) pushed this change at such enormous cost.

In Chapter 4, I discuss changes in the landside–airside boundary as a result of the world's massification of air travel and the new availability of planning experts. Beyond pressures from the market and the ever-changing technological conditions of the airport, the fact that airport planning became a well-established specialty in universities and technological institutes had a strong impact on the way airports were conceived. For example, at Tampa International Airport or TIA (1971), some of the first qualified airport planners pioneered the way the landside–airside boundary could be set in a different spatial dimension, which they called the "Landside Airside Concept" (1963). In this case, one of the leading airport-planning consultants in the country, Leigh Fisher and Associates (LF&A), and not the FAA, would persuade the officials of Hillsborough County to re-study the "quantifiable inefficiency" (LF&A, April 1963, p. 38) of current airport terminals. The outcome was a radical innovation that became an instant airport-planning touchstone and a model adopted worldwide. The landside was physically separated from the airside but linked through a "People Mover," shortening the long walks and avoiding the so-called delay. For the first time ever, buildings were referred to as "airside" or "landside." In the mind of Marge Brink Coridan, Tampa's head planner, airside included both the apron and the satellite buildings split from the center; landside referred to a central facility that would work as a transportation node. In fact, LF&A rapidly copyrighted the revolutionary scheme. This section collects and analyzes the key voices of those experts who intervened in the conception of TIA.

In Chapter 5, I develop two themes: First, I draw the line between airport reinventions and airport innovations, and thus help the reader by clearing up any blurriness on this matter created by discussing the previous case studies. Second, and foremost, I try to answer the following question: Are boundaries repeatable pieces of a mechanism, or are they autonomous entities, subject to the absorbing of cultural practices? In Sect. 1, "On Airport Reinventions," I discuss how congestion can be highlighted among the causes that prompted full airport reinventions. I also comment on the consequences of my previous three case studies: LaGuardia, Dulles, and Tampa. Section 2, "Airport Innovations," speaks of historical changes, departing from what is commonly known in airport planning as the standard, "linear model." I present two iconic cases in which airport layout ideas were introduced that were perceived as new: the 1974 Terminal 1 at Paris's Charles de Gaulle Airport and the 1960 Pan Am Terminal 2 at New York's Idlewild Airport (later known as JFK). Last, in Sect. 3 I argue that airports are not clones, or even repeatable copies, as many people may think. I support this claim by showing that the frontier between the airside and the landside captures many cultural differences that make each airport unique. These distinctions produce a mirror effect that reflects either local customs or the way these local forces reacted against foreign or global models. Toward that end I speak of three topics: The causes and legacies of airport reinventions, the process of landside–airside boundary innovation, and how these boundaries add transparency and allow us to glance at the cultural mirror. In addition, I address the differences between planning new airports and expanding them in phases, and finally, I compare boundaries at regional and international airports.

The increased flow of global passengers and the governmental obsession to control each one of them may help to explain why airports resemble liquid's containers and, how according to location and culture, they may feature certain degrees of *permeability*. In Chapter 6, I discuss the growing relevance of the landside–airside boundary as a fine-grained filter that controls the density of airport terminals, and eventually, the flows of those liquids that try to pass through. If it is true that passengers and employees need to flow through an airport like liquids—in addition to the infrastructure's natural *porosity*—designers who were increasingly pushed to look for separation and new strategies of containment started envisioning the airport as a system of control *valves*.

But are there any features that become visible only through the use of a landside–airside boundary as a lens? In trying to answer this simple question, I bumped into some interesting methodological speculations. First, the identification of these boundaries opens imaginary windows that may be described as technological "viewports." These portals allow us to see the in-between space, which is usually otherwise inaccessible and marks the territorial border between us and the hardware of technological systems. Second, I have also found that in order to make visible what has remained hidden within those contained boundaries, viewports should meet some conditions. At least in the case of airports, I have found at least five features that characterize these portals. Viewports should be discursive, normative, objects of knowledge, transitory, and intangible.

Each aspect has been developed separately: The boundary is *discursive* because, for example, it becomes a representation of the rise of international security concerns in the mid-1970s, and as internal and external need for security grows, so does the need to filter, sort, and split more efficiently and perhaps even more drastically. It is *normative*, because alongside the national-security discourse came a new set of rules that specifically placed the boundary at the center of the debate; the boundary is also an *object of knowledge* because in the context of the events of 9/11, a new literary production on airports arose, one much more committed to a social, political, and legal agenda that questions the very nature of the airport and its limits, past and future use of technologies, the regulatory frames, and the rising pressures of retail. The boundary is also *transitory* because, although wait time has greatly increased during recent decades, the "passing" nature of these spaces turns every effort to make it look permanent, as architecturally themed or artificial. These "stretched" frontiers have been described as areas of "*placelessness*." For retail purposes, this illusion of autonomy has been priceless. Airports have created micro-worlds in which people could even live, as in the case of Sir Alfred (see anecdote on Chapter 6). Last, boundaries are *intangible*, because as security requirements that had to be attached to—or often forced into—existing terminal buildings, new virtual technologies were introduced to compensate for those buildings' physical limitations (such as surveillance systems or scanning devices, for example). Because the construction of an immaterial borderline was ominously imminent after 9/11, new security measures came equipped with psychological constructs, the informatics frontier, and the establishment of non-material mechanisms of control.

Perhaps airports inherited from aircraft their aura of mystery. Because airports have grown as complex systems contingent on a larger system, as Douglas (1995) has argued before, they have developed a sort of augmented complexity that is often labyrinthine and hard to unthread. Since the early days of aviation, airplanes have disappeared every now and then for unexpected reasons, including flocks of birds, terrorism, and human error—or, as any with other technology, simply because they are subject to malfunctions and accidents. Often this happens in hard-to-reach areas of a diverse geography. Therefore, engineers of the post-Second World War era decided to build bright orange boxes—metaphorically called black boxes—which were able to record all human and plane activity as voice and data, in order to grasp the causality of incidents. Aircraft manufacturers usually feedback from these black boxes and inform their new designs and innovation processes. Perhaps Rosenberg (1994) would have labeled this idea "learning by using." In general, the timeline of technological change of aircraft has been characterized by secrecy and uncertainty—especially in the way things have been communicated to the public. This is due perhaps to the intimate relationship between aviation and war, and to the public and private (governmental agencies, contractors, and war-gear fabricants) idea of seeing planes more as weapons than as means of commercial transport.

During the last years, I have opened a mesmerizing conversation with some of the brightest historiographers of airports and aviation. Out of this meaningful and often mentorship relationship, I have learned how airports were invented and why, how people reacted to the magic of aviation, how distant worlds came together through the speed of airmail, and how modern airports became a reality. I owe much of this work to my candid talks with Debbie Douglas, Marc Dierikx, Peter Lyth, Joe Corn, Grace Torre, and Gordon Pririe. Other authors and historians such as Bednarek, Wohl, Gordon, Stoff, Holden, Szurovy, Zukowsky, Greif, Bosma, Brodherson, Bruemann, Stunkel, Lockhart, Rau, Voigt, Meisse, During, and Przychowski have also enlightened me with their personal depictions of airport history. From S&TS's body of literature, authors such as Vincenti (1976) have tried to show an intellectual, engineering culture responsible for developing aeronautics; Gallison (Gallison & Roland, 2000) has gone in-depth inside the mechanism of the production of flight technology; Weber (1997) has questioned women's access to military aircraft thanks to the design of physically gendered cockpits; and, more recently, Downer (2006) has unthreaded the secret testing of airplane turbines, shooting frozen turkeys at them with homemade cannons.

In S&TS, scholars adopted the concept of black boxes to describe the inputs and outputs of technological devices, or the "inner workings of which need not be known for it to be used" (Sismondo, 2009, p. 97). But rather than focus on the outputs, such as facts and objects, according to Sismondo (2009), "S&TS investigators pay particular attention to controversial stages in their histories" (p. 98). Thus, my study of the landside–airside boundary focuses precisely on those critical moments of controversy when actors found only a rhetorical closure to the problem of reinventing the airport. The three case studies analyzed further in the following pages share the connection of being historical successes and failures at the same time.

On the one hand, actors may have shown both frustration at not being able to reach a definitive new model to solve the airport (closure) and pride at being the minds behind newer airport models that gained a larger consensus, as in the case of Tampa's team of planners and designers (rhetorical closure). The world of airport planning in particular seems to be plagued with black boxes. As I mentioned before, the art of planning airports has shown an exponential level of complexity, usually spinning around the experts' race toward the perfect airport layout. Encouraged by this competition, planning firms have jealously vaulted expert knowledge. In the 1960s, Leigh Fisher in the USA and ADP in France (*Aéroports de Paris*) were positioned at the forefront of this race; under public financing, both firms researched and speculated on the new whereabouts of future airports, thus gathering enough theoretical and practical information to consolidate them as world leaders in airport design consultation. The idea of opening black boxes in airport planning is based on the initial assumption that most of the landside–airside negotiation processes have been archived for different reasons throughout time, rendering the boundary hardly visible to non-expert eyes and, furthermore, to the non-participant at these tables.

It was precisely when I managed to interlace both my own participation as a design professional in planning meetings (which made me aware of the relevance of landside–airside boundaries) and my first approach to some key S&TS concepts such as the principle of symmetry (or a fair account of histories in order to understand and open black-boxed knowledge), large technological systems, and boundary objects that I conceived of the idea of using landside–airside boundaries as socio-technical constructions.

Finally, it was when I witnessed myself how complicated and painful was the process of fixing a specific solution for a landside–airside boundary that I massaged the question of why, after a century of trial and error, we have not reached agreement on how an airport should be. Although I was already familiar with the history of airports, I decided to embark on a comprehensive revision of it, but this time looking for relevant cases of controversy or change precisely at the frontier between the landside and the airside. It was then that I suggested that if we use these boundaries as windows (or viewports, in the conclusion), we would be able to see airport history in a whole different way.

I maintain in this book that airport planning and design have moved along a *straight line* for more than a century. This pattern of *linearity* has been supported by what planners call "standard airport design," and it basically consists of learning from what is established[3] as modestly successful (airport typology) and introducing small, superficial changes to the selected scheme (finger, satellite, pier, unit, etc.). The historiography of the airport is most often pictured in terms of spectacular achievements, mostly architectural and spatial, but this interpretation could be deceptive and misleading. Why? Because most airports are based on the same standard airport layouts of the past (sometimes almost a century old). This "standard airport design" may be seen as a linear path of development, wherein some irregularities (social responses) have protruded, forming what Hughes defines as "reverse salients." Throughout the text, I make specific use of S&TS ideas and theoretical concepts, such as Nye's interpretation of the "technological sublime" in Chapter 2; in the case of Dulles Airport, I used Hughes's (1983) model of "large technological systems" and its concept of the "system builder." Throughout other sections, I use two of Akrich's (1992) concepts: the "heterogeneous technical object" and her idea of how technological designs become "imprinted." Last, in the case of Tampa's airport, I have used Pinch and Bijker's (1987) theoretical concept of the technological frame in order to understand the interactions among the relevant social groups. Finally, in SCOT's terms, I have argued that in all three reinventions of the airport, authors reached only a rhetorical closure. Throughout the chapters, I have also paraphrased some of Scott's (1998)

---

[3]"*Established*" by experts and administrators, not necessarily by users.

ideas about how the vertical powers of the state and other institutions prostrate social interests and so forth. As I explain in the last remarks of this book, finding the landside–airside boundary without an S&TS framework would have been a nearly impossible task.

In the present day, we often transit through the boundary's limits without realizing it; however, its psychological dimension is indeed present in any airport. This condition has complicated the identification of the landside–airside boundary and the challenge to trace it back into the past, even harder. But just a glimpse at the first years of aviation may evidence a radical departure from our tactile relationship with airplanes toward our present physical *divorce*. This condition clearly suggests the existence of a story that has not yet been told. Our right to enjoy the world of aircraft has been strangely denied, but maybe in the shadows of the war against terrorism this is not so unexpected. However, back in the first quarter of the twentieth century, the lack of a boundary between the landside and the airside captured the romantic days when an "airminded" public enjoyed visits to airfields just for mere entertainment, although the portrait of this remarkable moment in time may suggest the first clues in the understanding of a growing divide between the airside and landside. Sadly, in our days the airport terminal is a sort of Bentham's Panopticon,[4] where surveillance and tension are omnipresent, passengers are treated as potential criminals, and people are not allowed to make jokes or laugh. A century ago, however, airports were something radically different, and that is where we now turn the page.

[4]I am referring to Jeremy Bentham's prison model "the Panopticon," where prisoners could never know when they were being surveilled. Foucault described the implications as a crucial instrument of mental discipline in his classic *Discipline and Punish: The Birth of the Prison* (1975).

References

Publications

Akrich, M. (1992). The de-scription of technical objects. In W. Bjiker & J. Law (Eds.), *Shaping technologies*. Cambridge, MA: MIT Press.

Braun, H.-J. (1995). From airships to airbus. In W. M. Leary (Ed.), *Airships to airbus: The history of civil and commercial aviation*. Washington, DC: Smithsonian Institution.

Douglas, D. (1995). Airports as systems and systems of airports. In W. M. Leary (Ed.), *Airships to airbus: The history of civil and commercial aviation*. Washington, DC: Smithsonian Institution.

Downer, J. (2006). *The burden of proof: Regulating reliability in civil aviation*. (Doctoral dissertation). Cornell University, Ithaca, NY.

Froesch, C., & Prokosch, W. (1946). *Airport planning*. New York, NY: Wiley.

Gallison, P., & Roland, A. (2000). *Atmospheric flight in the twentieth century: Vol. 3. Archimedes new studies in the history and philosophy of science and technology*. New York, NY: Springer.

Hughes, T. P. (1983a). *The evolution of large technological systems: The social construction of technological systems*. Cambridge, MA: MIT Press.

Leigh Fisher Associates, Inc. (1963a, April). *Study of U.S. airport terminal building as a basis for planning new passenger terminal facilities at Tampa International Airport* (Vol. 1). San Francisco, CA: Author.

Perlman, H. (2004). *Airports: A century of architecture*. London, UK: Lawrence King.

Pinch, T., & Bijker, W. (1987). The social construction of facts and artifacts: Or how the sociology of science and technology might benefit each other. In W. E. Bijker, T. P. Hughes, & T. J. Pinch (Eds.), *The social construction of technological systems*. Cambridge, MA: MIT Press.

Rodrigue, J.-P., Comtois, C., & Slack, B. (2009). *The geography of transport systems* (2nd ed.). New York, NY: Routledge.

Rosenberg, N. (1982). *Inside the black box: Technology and economics*. Cambridge, UK and New York: Cambridge University Press.

Rosenberg, N. (1994). *Exploring the black box: Technology, economics and history*. Cambridge, UK and New York: Cambridge University Press.

Scott, J. C. (1998). *Seeing like a state: How certain schemes to improve the human condition have failed*. New Haven, CT and London, UK: Yale University Press.

Sismondo, S. (2009). *An introduction to science and technology studies*. Malden, MA: Blackwell.

Star, S., & Griesemer, J. (1989). Institutional ecology, "translations" and boundary objects: Amateurs and professionals in Berkeley's Museum of Vertebrate Zoology, 1907–39. *Social Studies of Science, 19*(3), 387–420.

Vincenti, W. G. (1976). *What engineers know and how they know it: Analytical studies of aeronautical history*. Baltimore, MD and London, UK: Johns Hopkins University Press.

Weber, R. (1997). Manufacturing gender in commercial and military cockpit design in science. *Technology & Human Values, 22*(2), 235–253.

Zukowsky, J. (Ed.). (1996). *Building for air travel: Architecture and design for commercial aviation*. Munich, Germany and New York, NY: Prestel and Chicago, IL: Art Institute of Chicago.

2

"The Romantic Borderline": From Fences to the Skywalk—Landside–Airside Space in an Early American Airport; New York LaGuardia Terminal 1933–1939

When the word "airport" was first used in 1919, there was no established body of knowledge or technological practices associated with airplane landing fields, no experts who claimed this work as their own, no "normal technology" as historian Edward Constant would put it.

—Deborah G. Douglas, "Spectacle of System: Making of the Modern Airport," unpublished paper presented at Society for the History of Technology Conference in Tacoma, WA, October 2010, p. 1

1 Introduction

Only after the Second World War were airports laid out the way they are now. However, between 1937 and 1944 a small number of projects helped to redefine our historical relationship with planes and with the fascinating world of aviation in general. Among these examples, at least two stand out: the 1937 Paris Le Bourget Airport's terminal expansion and the brand-new 1939 New York Municipal Airport, also called North Beach and years later better known as LaGuardia Airport (see Fig. 1). The French expansion was clearly a sound success and rapidly became a paradigm for how to grow

V. Marquez, *Landside | Airside*, https://doi.org/10.1007/978-981-13-3362-0_2

Fig. 1 Two-level split at LaGuardia Airport, separating for the first time departures and arrivals. A break though innovation that would make a huge impact in airports worldwide. *La Guardia Airport Passengers* [digital] (1944) (Reproduced with permission of Underwood Archives/via Getty Images)

existing buildings. Its architect, George Labro, imagined his building as a long pier alongside which aircraft might be comfortably parked. However, in terms of dealing with the landside–airside boundary, it retained the concept of the classic airport of the interwar years.

In contrast, New York's LaGuardia represented a radical departure from its contemporaries. Just as Moncure Robinson transferred back to England the first steeped railway[1] in 1867 (Stapleton, 1987). Americans sent back a new airport scheme to Europe and to the rest of the world. LaGuardia distinguished itself from the rest, perhaps because it was

[1]A steeped railway is an American innovation that literally tilts a railway track over an inclined plane.

the first of its kind and literally the first modern airport in America. Unanimously, its contemporaries considered it to be an exemplary technical innovation in airport functionality (airport planning did not yet exist), although its architectural aesthetics was always strongly criticized. As I suggest in this chapter, some specific features of this airport's layout captured, for the first time, the growing conflict between humans and machines, the irrevocable breaking-off of spectators from airplanes, an innovative layout that remained as a *silent witness* of an era, until the terminals were finally demolished in 1960 (Arend, 1979, p. 78). At LaGuardia, two innovations were evident: first, thanks to the spectacular Skywalk, visitors were split from passengers at the apron; second, for the first time ever, inbound and outbound passengers were separated on two levels. This very American solution came from the hand of the airport designer, the distinguished architect of the rich and famous, William Delano, who had great expertise in designing beautiful mansions and social clubs.

Delano privileged the need for an efficient way to drop off and pick up passengers with great comfort. Thus, he conceived the fancy idea of attaching an ample double-deck motor lobby, which he would call later the arrivals and departures curbside. Nonetheless, LaGuardia's breakthrough is only visible as a reaction against the early airport of the first third of the century. So why did airport terminals become necessary and why did they become congested so quickly?

2 Pilots Are Gods Descending from Heaven

Public perception of aviation during the first four decades of the twentieth century was diametrically opposed to our present views. In the earliest days, flying machines were an instant technological sensation. In this vein, Marcel Proust wrote in his *In Search of Lost Time*,

> I was as moved as a Greek would be who saw for the first time a *semi-god* ... The aviator seemed to hesitate in choosing his way; I felt there lay open before him—before me, if habit had not held me prisoner—all the routes of space, of life; he flew away, glided for a few instants above

> the sea, then brusquely making his decision, seeming to surrender to an attraction the opposite of that of gravity, as if returning to his homeland, with a light movement of his golden wings, he ascended straight up toward the sky. (Proust, 1954, p. 1029)

Proust's awe of the *supra* human character of the pilot poetically transforms into clairvoyance and reverence.

The unparalleled idea of navigating across the skies gave these men and women a mythical and almost ethereal proportion; the high risk involved before aviation became a stable system made spectators praise them as the bravest of people. The chronicles of war combat, acrobats, and unimaginable explorations gave them the status of heroes. The myth of the winged men forged a culture of reverence and unlimited veneration that even reached a metaphysical level. The machine was a means of reaching those "higher and deeper goals" that Nietzsche had placed above the Darwinian struggle for existence as a motive for human life and effort (Wohl, 2005, pp. 277–278). Yet the dialogue between spectators and airplanes represents only the temporary celebration of the sublime. For Nye,

> the technological sublime does not endorse human limitations; rather it manifests a split between those who understand and control machines and those who do not. In Kant's theory of the natural sublime every human being's imagination falters before the immensity of the absolutely great. In contrast, sublimation based on mechanical improvements is made possible by the superior imagination of an engineer or a technician who creates an object that overwhelms the imagination of the ordinary men. (1994, p. 60)

Controlling aircraft was not an easy task. The flying machines were always accompanied by the bravery of piloting and therefore strongly related to the implicit dangers of injury or death. In most cases, fame was inversely proportional to the masculinity of the accomplishment. In fact, most expeditions and trips were encouraged by public challenges and financed with juicy rewards offered only to the most intrepid. This became a common way to advertise, for example, when a New York restaurant sponsored the famous transatlantic crossing challenge in the 1910s.

During the first decades of the century, airplanes had a certain meaning for people. The flying machines elevated the imagination and awakened higher ideals for a better world. In this regard, beyond being considered as mere equipment, they represented a binomial relationship between "men and machines."[2] The plane and the pilot were inseparable parts of an equation; they signified the bridging into a new era of unlimited progress. Spectators idolized the object, using their five senses to appropriate some of its essence.[3]

It became common to see people staring at the sky in search of an aircraft or desperately trying to touch airplanes when they had landed or were parked at an airfield. The classic example is the massive reception of Charles Lindbergh at the Paris Le Bourget Airfield after his Atlantic crossing in 1927. Hundreds of thousands of people gathered at the apron, jumping at the plane while it was still in motion, eager to tear off a souvenir from the plane or the pilot himself. Before a tragedy could occur, the guards miraculously rescued Lindbergh (Wohl, 2005, p. 12). Throughout time people's love affair with aviation became, not surprisingly, a continuum. The worship of pilots as brave heroes started in tandem with the diffusion of aviation and continued up until the Second World War. Airfields were transformed into airports, but people still flocked to view the planes; the early dreams of a better world driven by the new possibility of flying did not stop but grew stronger up until the days of the 1939 World's Fair in New York City and the renewed dreams of "Futurama." At the fair, perhaps the most popular exhibition was GM's pavilion, designed by Norman Bel Geddes. Ten million people, an average of twenty-eight thousand daily, took the "magic, Aladdin-like" (Gordon, 2004, p. 116) ride that simulated an air trip across the USA in the year 1960 (a ride into the future, of course!).

---

[2]Hence, they experienced flight vicariously through forms of mass culture: best-selling books, newspapers, popular magazines, comic strips, radio, and, above all, motion pictures. Aviators, like sports figures and actors, became celebrities, and subsequently, the early history of aviation must be understood within the framework of the rise of a certain type of mass culture (Wohl, 2005, pp. 4–6).

[3]In terms of technological meaning and my own interpretation of Heidegger's "The Question Concerning Technology," the man-machine *reveals* to the audience the *supra* human, just as the *miracle* does to the devoted.

"Futurama was at that time the largest animated model ever made. After a long wait that sometimes extended to two hours as they made their way along serpentine ramps on the outside of the building, spectators were seated in luxurious six-feet-high armchairs contained within compartments that held two people. They were then moved along a conveyer of over a third of a mile through different levels of the building that gave the illusion of flying over the land and cityscapes of the future at low altitude" (Wohl, 2005, p. 295). "At various points across the countryside, Geddes placed dirigible stations and small, regional airports that were connected to cross- country train lines. In the middle of the continent was a circular airport with three giant elevators that carried airplanes down to subterranean hangars. The Futurama ride ended at a city of glass towers and a giant airport three miles in diameter. 'Its entire area is paved,' said the narrator's voice, making it possible for planes to land or take off in any direction and in large numbers" (Gordon, 2004, p. 116). The door that aviation was opening became a great source of inspiration and debate for many thinkers of the epoch. Visionary projects by planners and architects such as Le Corbusier, Antonio Sant'Elia, and Tony Garnier, and several airport competitions such as the one sponsored in 1929 by Lehigh Portland Cement, urged the development of the airport's integration with the city (Brodherson, 1996, p. 68). "The airplane, in Le Corbusier's mind, represented clearness of function and the daring required to break with the past. New machines, new men. They [the aviators] are filled with enthusiasm, the pleasures of daring, of breaking with current stupidities … In aviation everything is scrapped in a year. Le Corbusier accompanied bold statements like these with an ambitious program for rebuilding many of the world's major cities" (Wohl, 2005, pp. 294–301). In his radical view, Le Corbusier envisioned great urban centers that were openly subordinated to vehicles, almost thought entirely for them, and, most particularly, for the aircraft. His designs relied so much on the new technologies that it made them look like scenes of science fiction; hence, most of the projects remained unbuilt and were often seen as unbounded and delirious.

As seen in competition schemes and panels, the proposals ranged from utopian floating airports in front of Manhattan or sets of skyscrapers linked at the top with aerodrome-like runways to the more grounded but still eccentric proposals of incredibly long buildings equipped with runways on the top floor. Others were slightly more pragmatic, with monumental roundabouts for aircraft takeoffs and landings in any direction or "drive-in"-like arrangements. In a few cases, the landside–airside boundary was addressed and designers envisioned the first facilities with arrival and departure differentiation in levels. However, this innovation remained on paper until the construction of LaGuardia. For a number of reasons, "even the least ambitious of these structural and architectural proposals to meet the airport needs of the metropolis, was never realized. Too many obstacles stood between vision and reality. For one thing, even in 1930 urban dwellers were sensitive to the noise of aircraft and would have protested mightily were such landing facilities ever built in their vicinity. That they did not reflect economics was a second serious problem" (Corn, 1983, p. 105).

At the birth of commercial aviation, pilots and aircraft were considered an indivisible entity. Particularly in mass media, the echoes of this link resounded up until the 1950s. In an implicit manner, this relationship was established on the premise of having no boundaries between men and machine. Our "affair" with aviation was an open contract with endless possibilities, and a good part of the public had faith in the platform of the future and relied on its future developments.

3 The Fence

If "wherever an airplane takes off or lands, there is an airfield," what is an airfield? (Douglas, 1995, p. 55). During the first decades of the century, landing fields were developed on grassed parcels or hardened runways equipped with simple buildings and maybe a few wooden or iron hangars. These flat strips of land were usually found on the fringes of urban areas or in the countryside and, according to Clack (1998), often were "laid out in the form of a Wind Rose" (p. 3). Airfields "handled only a few small planes each day. Airline companies had their headquarters downtown, where the few passengers who traveled by air assembled for their journey. They were taken to the airfields by a special car. They walked across wet grass to reach their plane, shook hands with the pilot, then as air travelers-became heroes of a new age of mankind" (Chadeau, 1995, p. 17).

As we see in this quotation, it was clear that no boundary separated passengers or visitors from aircraft. This intimate relationship materialized in the concept of "aerodromes," where planes performed acrobatics for spectators. Like horse races in hippodromes, air races were organized and several airfields were transformed into aerodromes equipped with platforms and fancy restaurants for the upper class. In tracing the origins of airports, Voigt (1996) has suggested that the "architectural language was rooted in the world of sports and racing, with airplane hangars modest wooden buildings replacing the racing stables. The terms aerodrome and airdrome, used early on in the history of airports, owe their creation to their relationship with hippodrome and its more recent variant, *autodrome* (coined about 1900)" (p. 28). Nonetheless, the democratic spirit prevailed, and the general public remained welcome to draw closer to the aircraft. These policies speak of the existence of an unwritten "right to access airplanes" that made aviation available to many, at least as a spectacle. Thanks to this way of thinking, families visited airfields on weekends not merely to be entertained, but to be awe-inspired. I think these facts are particularly interesting to an understanding of why aviation became so popular, although airfields became the favorite place to go, mainly for white people.

Photographs of the time capture the earliest version of the landside–airside boundary. Just as in hippodromes, a fence divided planes and their airfield from cars and their parking areas. The haphazard flow of vehicles at the apron or even at the runway became a growing danger. For the sake of safety, two types of fences were typically installed. The first was usually a wire mesh fence separating automobiles, as parking areas were in full proximity to the apron. The second was a chain fence, clearly less formal, which prevented the invasion of the apron by large crowds during special events. Interestingly, some of these images also show special access doors and download areas for vehicles transporting luggage and food; these sections were also demarcated by railings, as in the case of Roosevelt Field in Long Island, NY, in 1928 (see www.landsideairside.com) or many other airports across the country.

Fig. 2 A key example of the early landside–airside boundary. Spectators are separated and contained in fenced areas. Leslie R. Jones, Photographer (ca. 1930) (Courtesy of the Boston Public Library, Leslie Jones Collection)

In a 1930s photograph of East Boston Airport (see Fig. 2), people crowd the apron but are *contained* by iron fences forming corrals. Access is limited through a number of buildings, and only passengers, aviators, and service people are next to the aircraft. However, some planes are parked within the corrals, and people are inspecting them closely or even touching them. But would airplanes and airfields ever be part of our everyday lives? As Corn (1983) commented, "Of all the predictions Americans made about the airplane, most of them expressed that sooner or later everybody would fly. Indeed, the expectation that democracy would prevail in the sky and that flying would become as common as riding or even walking was the cornerstone of the winged-gospel. Without universal flight, prophets could not imagine a transformed society. In the coming air age, therefore, people would take to the air, it was predicted, not merely as a means of travel but for their health, for recreation, and even to perform their jobs" (p. 32).

Charlotte Perkins Gilman, the writer and radical theorist of the turn-of-the-century women's movement, even posited the emergence of an "aerial" person. In an article in *Harper's* titled "When We Fly" (Perkins, 1907, p. 1664), she predicted that the new type would possess a wholly different value system. Unlike inferior "earthy" man, the new "aerial" man "cannot think of himself further as a worm of the dust, but [only] as butterfly, psyche, the risen soul." The new attitude would bring about human "intercourse on a new plane," she predicted buoyantly. For nearly half a century preceding the *Popular Mechanics* article cited by Corn, "millions of Americans expected to own a flying machine in the near future. Like most Americans who gazed into their crystal balls to glimpse the future air age, these prophets were not excited just because people would fly tomorrow. It was rather the way of life that flying would make possible that sent them into rapture. Instead of 'rushing like moles' into dark subways and breathing 'foul subterranean air,' wrote a 1922 enthusiast, tomorrow's sky commuter would escape the 'push and struggle' with the urban 'masses' and whisk swiftly and pleasurably through the clean, clear air alone or with a few friends. Inevitably, prophets envisioned the coming aerial lifestyle as a rural one. The airplane 'would create countryside,' just as the automobile had created the suburb. As the comparison to suburb suggested, air-car prophets envisioned the further spread of commuting. The future would see thousands of city workers air day in 'rising like homing birds' and flying off 'to cool mountain retreats'" (Corn, 1983, p. 91). Aircraft would become the means not only of linking rural residences and the urban workplace but also of shopping, making social calls, and taking vacations. Rather than once-a-year outings to favorite lakeside or mountain retreats, waxed an advertisement during the Second World War, the "family car of the air" would facilitate "vacations every week end" and "magic trips to everywhere" (Corn, 1983, pp. 91–92).

Hence, in the case of the relationship between people and aviation, the diffusion of the innovation spread across the country and helped to spread the idea that airplanes would have the same reach automobiles did, and therefore, the dream of having an airplane in every house would become a reality. In that sense, both urban and rural societies would finally be modernized. These ideas were clearly based on the widespread assumption that "autonomous technological forces drive social change" and in this regard are quite similar to the culture of the automobile. The dream of a motorized age was then transforming into a dream of an air age. Kline and Pinch's (1996) essay on the "social construction" of the automobile sheds light on how diverse social groups, especially in rural areas, became strong "agents" that literally helped to reshape these emerging technologies (p. 764). In the case of aviation, the modernizing efforts on the home scale became futile in both city and rural environments for a simple technical reason: airplane technology required a runway of considerable length for takeoffs and landings. In this sense, rural America had an advantage—the availability of space—but this condition resulted in a new paradox: Aircraft were never mass produced for that same reason (runways) and therefore remained expensive and inaccessible. However, during the 1930s and with the birth of commercial aviation, the idea of building an airport for every town grew stronger.

4 The Romantic Borderline

In this section, my interest is in discussing the birth of an airport as a concept, the early stages in the formation of systems of airports, and how they were funded. These aspects are relevant to understanding the forces that shaped the *modern airport*. In this regard, I will comment on the identification of the agents of change, particularly on how the increasing demands of new aircraft technology would become critical aspects of the *soft transition* of the landside–airside divide. But in this sense the "idealized 'tomorrow' never comes. The airport is at once a place, a system, a cultural artifact that brings us face-to-face with the advantages as well as the frustrations of modernity" (Gordon, 2004, pp. 4–5).

It is commonly believed that the military played the leading role in the creation of the "modern" airport; new studies of the social history of technology, however, contradict this. Well-documented research shows that it was in fact the US postal system, along with the rapid establishment of commercial aviation, that defined the future of airports in the early 1930s. However, "airfields had been built all over the world with military funds" (Higham, 1995, p. 21). In fact, just after the war, "commercial aviation—the transport of paying passengers in converted bombers, was born" (Toulier, 2000, p. 12), and with it the wide popularization of air travel. Many of the terms commonly used to describe *airports* have their origins in the iconology of maritime, rail and race-track facilities. Voigt (1996) has written that "just as *aircrafts* derived from the iconology of sailing, which had shaped the idea of manned flight as early as the seventeenth century, when a writer such as Francesco Lana di Terzi dreamed up imaginary *airships*, in which bold *aeronauts* sailed across the *sea of air*" (p. 27). In a relatively short period, the literature reflects the use of interchangeable terms; *landing fields* or *airfields* became *aerodromes* and later *air-ports*; and even the distinct term *air-terminals* appeared. Interestingly, the military did play a critical role in this differentiation. Air Service drew a "distinction between simple *landing fields* and what they called *airdromes*" (Bednarek, 2001, pp. 17, 19, 73). There is no clear indicator in the literature, but one could logically assume that the Air Service favored that word as a result of its wartime experience in Europe, where it was one of many terms used to describe aviation facilities. Surprisingly, relevant figures like Samuel Langley, a member of the Smithsonian and a real pioneer of American aviation, used to call them *aerodromes.*

But not everyone agreed on the terminologies. For B. Russell Shaw, interchangeable names were often assigned incorrectly: "A landing field was an area designed to, or simply large and flat-enough to allow the arrival and departure of airplanes. An *airdrome* was a more fully developed facility including hangars and refueling equipment in addition to an adequate landing area" (Bednarek, 2001, pp. 11, 18, 19). In his opinion, a *landing field* was "a tract of land comprising less than 100 acres, level but without accommodations except small hangars, gasoline servicing facilities, telephones, pilots' quarters, rest rooms and a passenger station" (Bednarek, 2001, p. 73). According to Shaw, an *airport* also provided "night lighting, first aid, maintenance equipment, fire fighting apparatus," and other equipment. An *air terminal* was a facility that had the highest quality airport equipment for established airlines. In fact, "it would match in quality and comfort the best railroad terminal facilities in the nation's largest cities" (Bednarek, 2001, p. 73).

By the end of the interwar period an *airport* turned into a site with multiple acting and interacting technological subsystems. But it was not until two decades after the Second World War "that both the facility we have come to know as an '*airport*' and a national '*system of airports*' was developed" (Douglas, 1995, p. 55). With the exception of Deborah Douglas and a few other authors, airports have not been studied as large technological systems. Douglas (1995) suggests that in addition to Thomas Hughes's classic model, Chandlers's modern industrial enterprise model should be incorporated and viewed as *complementary.* Airports are technological devices just as aircraft are, although on a different scale. In addition, they are complex systems themselves and when they interact with other facilities of their kind form networks of airports. Historian of science Rachel Laudan maintains that "the distinction between devices and systems, while useful for analysis, is to some degree arbitrary; a complex technology can be often regarded as a device" (quoted in Vincenti, 1976, p. 201). When airports became regulated by the CAA, they were still notably disarticulated in between. Following Hughes's (1983, pp. 6–17) technological systems model, in terms of regulation, financing, and engineering problem-solving, airports seem to be still a bit far from constituting systems.

During the 1920s and early 1930s, the government created a number of commissions that gradually issued general regulations for airport construction. In addition, the Air Service pioneered the promotion of the establishment of municipal landing fields and also offered cities "advice on where fields should be located, their size and shape, the character of the ground, the approaches, markings, and accommodation" (Bednarek, 2001, p. 34). In the name of safety, these federal commissions started regulating planning issues in order to give uniformity to the process of airport construction. Among these initiatives, none was as influential as the point system used by the Secretary of Commerce's Air Branch. Airports would range from A to D, from the best equipped to the most basic. Counties across the country felt the need to *modernize* and build their own airports (to increase commerce, perhaps), although resources were limited.[4] In this regard, Franklin D. Roosevelt's arrival as president sped up the allocation of federal funds for airport development. According to Corn, President Roosevelt "understood the importance of aviation as being a practical as well as symbolic necessity in the nation's recovery. He established the Civil Works Administration (CWA), and among its expenditures were $11.5 million for airport construction. Federal support increased further in 1935 when the Work Projects Administration (WPA[5]) came into being and Division of Airways and Airports was established under the WPA's chief engineer" (Gordon, 2004, p. 98). (Roosevelt's more intimate environment was highly influential in drawing his interest toward aviation.)[6]

The 1930s also saw the introduction of an important new airplane and a change in the rules of the airmail game. In 1933, the Douglas Aircraft Company developed the DC-1 as competition for the newly introduced Boeing 247 (Bednarek, 2001, p. 98). As suggested before, the new concept of airport is clearly based on the increasing complexity of its *functions*. The understanding of airports as systems and also as part of larger systems increased the interest in *standardization of features* and *routines*. The major concerns in this regard were to improve safety, to maximize efficiency (technical), and to *speed up processes* (commercial). In this section, I attempt to answer this question in terms of technological innovation: Did new airplanes *really* demand new airfields? For Bednarek, new aircraft played a critical role in the ways airmail was organized. According to the author of *American Airports*, "the introduction of an important new airplane and a change in the rules of the airmail game marked a milestone in technological advancement. In 1933 the Douglas Aircraft Company developed the DC-1 as competition for the newly introduced Boeing 247" (Bednarek, 2001, pp. 98). The open competition between the two models sparked off the rapid development

[4]Paraphrasing Bednarek (2001, p. 40).

[5]According to Bednarek (2001), "help first came in the form of federal work relief programs. Federal aid thus provided airport with manpower but little in the way of materials or the latest equipment. Critics charged that mandating a link between airport aid and work relief limited the value of the program" (p. 98).

[6]According to Corn (1983), Mrs. Roosevelt showed an unbounded enthusiasm for aviation that fell in the "airmindedness" of the epoch (p. 57).

of aviation technology. As planes became larger and faster, new technological demands came up.

The US airmail company system was by far the largest user of air services in the country. Only one year after the launch of the DC-1, according to Bednarek, Postmaster General James Farley convened a meeting for the newly reorganized airlines, and the bidding for routes began anew. With new contracts and new types of aircraft, airport technology also underwent a period of innovation and development. A "dramatic increase in size and range of landplanes won the day" (Higham, 1995, p. 21). In 1934, the DC-3 was at least twice as heavy as the majority of its competitors. Heavier aircraft required longer runways with a more stable surface than sod or cinders. The growth of passenger traffic (from 500,000 in 1933 to 1.2 million in 1938) put enormous pressures on existing terminal facilities. Consequently, an increased number of passengers required more efficient transfers between air and ground transport. New airports required all operations to be scaled up to accommodate the new demands of commercial air transport. The fact that aircraft innovation produced increased technical ramifications was a consequence of the growth of aviation as a system. (It is important to remember that during these years, there were no hubs, and airlines have monopolized certain routes.)

Now, the *romantic borderline* was in danger. As aircraft became more sophisticated, the airport devices became a technology of their own. The simplicity of the aerodrome's *fence* was under evident pressure when the rise of the new airport technology "now necessitated careful studies of traffic circulation and ground movement of aircraft" (Bednarek, 2001, pp. 68). New designs were proposed that featured parallel hard-surfaced runways allowing simultaneous takeoff and landing operations. Zoning laws became much stricter as airport managers became increasingly vigilant of any obstructions.

The premonition of a gloomy future was embedded in a simple, technological maxim: "The heavier the aircraft, the longer the runway," and this idea affected assumptions about airport size and location. "Planners argued that additional land should be purchased and held in reserve" (Douglas, 1995, pp. 69). Engineers took this relationship too seriously; thus, airports first responded to aircraft innovation. They responded to these requirements, and in doing so, they started to break the implicit pact between people and *their* aircraft; the days of the flying man-machine were numbered.

Just a few years ago, Robert van der Linden[7] brought to light an alternate explanation for the conventional historiography of early aviation and airport transformations in the decade of the 1930s. Traditionally, historians of aviation have credited the efforts of the FDR Administration "to bring order to chaos" thanks to the establishment of a number of commissions, agencies, and regulatory bodies at the federal level. Although van der Linden (2002) credits this fact in part, he builds a counter-argument (pp. xi–xii). He regards Herbert Hoover as a pioneer in promoting the aviation industry, thanks to the establishment of the economic incentives of airmail's subcontracting. He focuses especially on General Postmaster W. Fogle Brown's role as the instigator behind the creation of a complex network of air routes and his business instinct to promote the newborn aviation companies. For others (e.g., Gordon, 2004, p. 28), Hoover's legacy for Roosevelt consisted of a number of unregulated monopolies such as the same US airmail subcontractors. This had a high cost for President Roosevelt, as history will later tell. A new generation had made fortunes from monopolizing aviation after the war, with great thanks to the unabashed favoritism of the Postmaster General. For example, the Navy pilot (Urwin, 2002, p. 27) and Yale graduate, Juan Trippe[8] was only twenty-six years old when he assumed control of Pan American Airways in 1925 and confessed to his friends his aspiration to become the JP Morgan of aviation (Gordon, 2004, p. 30).

The strong presence of the postal service is also evident in the imagery of the time. I have been particularly interested in observing historical photographs and airline promotional posters. For example, United Airways publicized its technological vanguard with a striking photo of a 247 Boeing flying over a caravan of horse-pulled carts

[7]Robert van der Linden is the curator of Air Transportation at the Smithsonian Institution's National Air and Space Museum.

[8]As I will explain in the next section, Mr. Trippe plays an interesting role in this study.

on the prairie. Its official purpose was to show the increased speed in mail delivery, but its underlying meaning was clearly to show the divide between two new worlds: the aerial and the terrestrial, the wealthy and the poor, the *modern* and the *obsolete*. In other photographs of the same epoch, we can see passengers board aircraft labeled "US Mail Express," or 1930s tickets and menus printed on postal paper and envelopes (including the characteristic blue-red rivet along the edge) (Garvey & Fisher, 2002, pp. 194, 40, 35). In addition, poster iconography was incredibly rich and complex. Several designs of those days show the strong link between the aerial mail and commercial aviation; some posters advertise air travel but at the same time include tables with estimated delivery times for correspondence (Szurovy, 2002, p. 66).

Back in 1928, Hoover's commission favored Newark's airport as the best alternative for New York City, causing a constant flow of resources to solidify its infrastructure through the years. By 1933, Newark was well established as the transcontinental hub for US airmail. Although the facilities were still rooted in the tradition of airfields, its solid profitability and the flow of federal funds kept it relatively up to date and meeting all post office requirements. As the trend of its operations forked into airmail and commercial airlines, Newark focused on promoting itself as the airport of New York City. Unexpectedly in November of 1933, as an airplane was approaching Newark airport a passenger loudly complained, arguing that his ticket clearly said arriving at New York and not New Jersey. Fearing a lawsuit, the pilot continued the trip up to Bennett Field in Brooklyn. The passenger was no less than the Major himself, Fiorello LaGuardia. LaGuardia was called "Major" for two reasons: The first was because he was promoted from Captain to Major after his contributions as a US pilot in combat on the Italian front in the First World War (Manners, 1976, p. 62); the second was because he was elected Mayor of New York City in 1933. His public "airmindedness" or, in other words, his unbounded passion for aviation, overcame the fact that he was injured "after a crash in his first solo flight" (Bayor, 1993, p. 38). Even after suffering this accident, he never stopped loving aircraft.

One of LaGuardia's greatest challenges was to bring New York into the air age: Aviation is established, he said. Nothing can stop it. The city needed an airport that was worthy of this importance. "And for this he had a loyal friend in the White House. 'Our Mayor is the most appealing man I know,' said President Roosevelt. 'He comes to Washington and tells me a sad story. The tears run down my cheeks and the tears run down his cheeks and the first thing I know he's wangled another $50 million'" (Gordon, 2004, p. 107). (I shall remind the reader that a portrait of LaGuardia's political ability is key in understanding what comes next, which is the realization of his dreams.) LaGuardia offered to invest money in Bennett Field. But there was no prospect in imagining a remodeled airport without the support of the US Post Mail service, and even with presidential support, it was a mammoth task. Post officers declared the idea of moving to another airport economically unviable, and without them, air carriers would not move either. The battle was led by the Mayor himself and took place on different fronts between 1933 and 1935. The biggest debate was proximity to the city. In March 1936, Postmaster General James Farley announced his decision. Airmail operations, because of easier access to Manhattan and the costs involved in the transfer, would remain at Newark and would not result in improved or less expensive service. This was a big setback for the Mayor's initiative (Bednarek, 2001, p. 109).

LaGuardia's next move resembles a classic chess strategy. When the challenger (New Jersey Airport Authority) was prematurely celebrating this defeat, the Mayor counterattacked by launching the construction of a full new airport in Queens. Thus, Newark's turned out to be a Pyrrhic victory. New Jersey politicians were scandalized. As technology was moving fast, they knew the disadvantages of competing against a state-of-the-art facility. It would clearly be a losing battle. New York City ensured an initial funding from the WPA, which eventually increased to more than $13 million, making it the first airport in the country to receive such a level of financing. Even without the support of Farley, LaGuardia managed to convince three of the four major airlines operating at Newark to move into the new airport. In 1938, the Civil Aeronautics Authority (CAA) held hearings and voted in favor of the requests by American Airlines, TWA, and United Air Lines to relocate to North Beach. One factor had key importance in this decision: the beach. Even against the advice of city planners, LaGuardia and his advisors chose the tiny airfield of North Beach[9] and the contiguous garbage dump on Riker's Island, simply because it had access to water. The airlines clearly foresaw the impact of operating Clipper airplanes[10] (luxurious flying boats) in trans-oceanic routes from New York.

From the conceptual stage, it was advertised as a landside–seaside facility. Strangely, this fact has been overlooked by airport historians and is rarely mentioned in the literature.[11] Two other good reasons to convince not just the airlines but the WPA itself were that North Beach was indeed twenty minutes closer to Manhattan than Newark and that reliable surveys showed that more than half the passengers at Newark originated in New York City. According to the Secretary of Commerce's Air Branch policy,[12] each municipality was entitled to have its own airport, so why not New York?

5 Reinventing an Airport in 1937: "Country Club" or "Rail Station"?

This section comments on two specific aspects: The first is devoted to having a better understanding of the designers, engineers and experts behind LaGuardia's airport; and the sources, the references and the organization of work. The second aspect discusses the technological transfer from rail technology and rail stations. Both of these aspects form the basis for understanding the final design and materialization of New York's Municipal Airport and what I later call "the soft transition" between the landside and the airside.

After presidential approval, LaGuardia vowed that the new airport would be ready for use for the upcoming World's Fair of 1939. Planning

[9]Formerly a mini Coney Island, North Beach became a favorite private land field serving the wealthy Long Island sport flyers, called Glenn Curtiss Airfield.

[10]The name "clipper" was borrowed from fast sailing boats from the nineteenth century.

[11]Maybe with the exception of Geoffrey Arend's brief comments on page 10 of his book *LaGuardia 1939–1979* in the collection *Air World Greatest Airports* (1979).

[12]Paraphrasing. Air Branch policy is fully discussed on page 46 of Bednarek's *America's Airports* (2001).

a modern airport without knowing exactly what that meant was already a serious challenge, along with such a close deadline. Although it was a joint effort of the municipal and federal levels, the airport planning fell on the shoulders of the WPA. Perhaps LaGuardia's promise "to bring New York's crumbling infrastructure up-to-date to build new highways, bridges and parks" (WPA, July 1, 1939b, p. 1) made him surround himself with the finest of military engineers. And certainly Lieutenant Colonel Brehon B. Somervell was one of them. The Works Projects Administrator was a prominent soldier, a radical conservative,[13] a West Point graduate, and a well-known businessman with proven political abilities.[14] According to the requirement spatial program (New York Municipal Airport, 1938), two facilities were needed: a *landplane* terminal and a *seaplane* terminal. The landplane referred to the airport itself, and the seaplane to the maritime terminal for transatlantic clippers. (It is interesting to notice that great emphasis was always put on the airport as a *terminal*, as I will expand later.) Surprisingly, Colonel Somervell established the division of work in not two but three large fronts.

The first front was relative to what the WPA called the *field* or the *field side*, which covered the *air* functions such as the airport layout; the planning of runways, lightings, and signals; the hangars and all the related civil infrastructure projects such as the site infill and the sanitation of the garbage dump. In this regard, Somervell declared that in "laying out the airport the best experience in the country in this branch of engineering was called upon" (WPA, July 1, 1939b, p. 2).

(In this quote, I find revealing the fact that the WPA did not employ the word *experts*, as if the *aeronautical consultants* of the time did not have enough experience in laying out airports.) The design of the field was the task of the Bureau of Air Commerce, but contributions from

[13]Colonel Brehon B. Somervell's anticommunist feelings sent the "History of Aviation" mural in Floyd Bennett Field to the garbage, simply because several characters appeared to be "strangely un-American in expression and thick mustaches that made them look like Joseph Stalin" (Gordon, 2004, p. 99).

[14]Somervell later became a key military figure during the Second World War. He took part in the Manhattan Project, was a member of the Army Corps of Engineers and has been credited as the author who conceived the idea of building the Pentagon.

the US Army and engineers of the airlines were also credited. The construction plans, though, were drawn by the WPA Planning Department and overseen by engineers from Parsons, Klapp, Brinckerhoff, and Douglas, and a general from the US Army.

The second front was referred to as the *seaside* and included the planning of the docks and hangars corresponding to the seaplane "port." For this purpose, the cooperation of Jon McKenzie, Commissioner, and Joseph Meehan, Chief Engineer from the Department of Docks was required. Additional engineering was necessary to design the system that hauled up the Clippers onto rail tracks, aided by dollies and electric winches. Both fronts were the charge of engineers and aviation *specialists.*

The last front was called the *landside* and was the responsibility of the architectural firm Delano & Aldrich of New York (see www.landside-airside.com). This front included the design of the buildings, including the *landplane terminal* and the *marine terminal*, the road system, and the landscaped and parking areas. In addition, Alexander Crosett was in charge of the structural design, A. F. Brinckerhoff of the landscape, and Syska & Hennessy of the mechanical engineering. Somervell's distribution of the work scheme would be enormously influential in the future of airport design development (New York Municipal Airport, 1938, pp. 3, 4). The divide not only separated engineers and specialists from architects and planners but also, perhaps indirectly, split the needs of aircraft from the needs of passengers and visitors. The airport was not designed by one mastermind, but was a negotiated solution along the way. Fronts one and two were well integrated by the WPA, but different interests meant that front three began to act independently.

The reasons for hiring Delano & Aldrich have not been made public, or at least not in the specialized literature of airports. Comments are usually a reiteration of the official version: Delano & Aldrich's firm was contracted because they had six airports in their portfolio (five, according to other publications). I find this explanation rather shallow and not necessarily realistic, and for the purpose of this chapter, this is a relevant matter. Hence, further investigations into Delano's personal correspondence and the firm's archives, as well as interconnections with LaGuardia's archive and other links, allow a better understanding of this decision. William Delano was a wealthy son of a New York banker, Eugene Delano.

Delano attended Yale College and later graduated as an architect from Columbia University. After graduating, he spent five years in Paris and received a diploma from the École des Beaux Arts. Like many of his contemporaries, Delano became a classicist. In 1927, he started a parallel political career as a consultant to the US Treasury Department. Two years later, at President Hoover's request, he became the architectural representative of the National Park Planning Commission and held that position until 1946, when he resigned.[15] During his career, he became an important ally and mediator between France and the USA.

William Delano's lifetime achievement as a "Beaux Arts" designer was notably rich and prolific, ranging from a large portfolio of exquisite residences (Cortissoz, 1924) and private clubs to the American embassy in Paris, the Military Academy at West Point, and Willard Straight Hall at Cornell University. His neo-classical palazzos across the Rhode Island gold coast included clients such as the Rockefellers, the Vanderbilts, and the Mellons. However, none of these merits seemed to me to be significant enough to win him the airport commission, so further inquiry was necessary. Delano grew up surrounded by a very influential social circle. In his early youth, President Theodore Roosevelt described him as a guy with a "little mustache and genial smile," as the Roosevelts and the Delanos were neighbors in Orange Mountain. His contact with power was ever present throughout his life (Pennoyer & Walker, 2003, p. 14).

During Hoover's Administration, William's cousin Frederic Delano was a member of the Committee on the Regional Plan of New York and its Environments. The main task of the group was to study the airport needs for the New York City metropolitan area (Bednarek & Bednarek, 2003, p. 109). Besides being relatives, Frederic and William were good friends and used to spend vacations at the Delano Hotel in Miami.[16]

[15] From the biographical notes on William Delano at the Archives of Yale University Library.

[16] Letters between W. Delano and Fred Delano from 1937 to 1941 William Delano Collection. The Archives at Yale University Library.

In addition, William's close friendship with the acting Secretary of the Navy, Charles Edison (who warmly referred to him as "Billy"),[17] won him numerous recommendations among politicians of the time. In those years, Delano received a major commission that linked him with Postmaster W. Fogle Brown and James Farley; it was nothing less than the enormous US post mail headquarters in Washington, DC. Considering the strong influence of the US postal service in airport planning (particularly in the NY/NJ region), this professional bond becomes intriguing.

In 1928 Juan Trippe, founder and chairman of Pan American Airways System, "needed to make a statement about his corporate aspirations" (Szurovy, 2003, p. 62). The new terminal building in Miami needed to meet the highest standards, as his airline was characterized as offering exotic destinations to the wealthy. Delano & Aldrich was the perfect choice: As Delano was a socialite himself (Eggebeen, 2007, p. 119), his portfolio was well-regarded by the East Coast elites. Plates of the time show the elegant detailing of his work and the upscale crowd who used it. This was the first of four projects developed for Mr. Trippe (Gordon, 2004). They included the peculiar, *country club*—looking marine Terminal at Dinner Key, which was intended to connect to the Caribbean and South America. My guess is that Mr. Trippe was happily impressed with the elegant private clubs previously designed by the firm, and in those days, the *terminal* was seen more as a *clubhouse*, just as private aviation was more a fancy sport than a part of the transportation industry. The accumulated experience of the firm working for Pan Am became a new element of consideration.[18] Last on the list of possible factors weighed in selecting them as the office responsible for the project is Delano's relationship to President Roosevelt. They were first cousins; William was eight years older than FDR. But both the

[17]Letters between W. Delano and Lt. Charles Edison from 1938 to 1940 William Delano Collection. The Archives at Yale University Library.

[18]Three more airport facilities were designed for Pan Am—Midway, Wake, and Guam Island—but these were just minor utilitarian buildings, prefabricated and exported from the USA (Pennoyer & Walker, 2003).

President and Delano rejected the notion that the family relationship had anything to do with the assignment of the job. As we can see in his private correspondence, the architect was certainly a prominent figure in the social scene of the time, and all the above facts speak to that. Nevertheless, it was said in the *New York Times* that Fiorello LaGuardia was favorably impressed with Delano's work in Dinners Key Sea-base, and it is highly probable that this fact also influenced his judgment (Pennoyer & Walker, 2003, p. 70). At Delano & Aldrich, Chester Holmes Aldrich, also a Yale and an ex-Beaux Arts student, was in charge of the office management and organization. As the firm continued to grow in the 1920s, they invited George A. Licht (Pennoyer & Walker, 2003, fn. 35) (owner of a small practice, Licht & Waterbury's Sons)—a promising young architect—to join the firm. He was regarded as an outstanding student and designer, having among his remarkable personal achievements the first prestigious Paris Prize in 1904, the gold medal in the Architectural League of America in 1907, and eleven medals during his stay at the *École des Beaux Arts* in Paris ("The architectural convention ends," 1907).

When the firm secured the airport commission, Delano associated with George A. Licht for the first time.[19] In architectural practice, it is common for large firms such as Delano & Aldrich to distribute the workload among its associates. These "associations" imply a higher degree of responsibility and therefore a higher level of creative recognition. Licht's associate degree in the LaGuardia project cannot be overlooked, particularly given that he was a talented designer. I later comment on his role during the design process, in which he was a "strong presence," according to his colleagues.[20]

Delano was brought up in a world where frequent travel to Europe was considered common. His contact with the latest airports was evident (Licht's as well), particularly in his frequent trips to France. In 1936, Le Bourget was just slightly different from what had welcomed

[19]As shown in most of the project documentation, where he is credited as associate architect. Columbia doc.

[20]As stated in a recent interview. David Aldrich, a former alumnus, in Pennoyer and Walker (2003, fn. 35).

Lindbergh a decade ago, a well-developed *airfield*. However, Paris aspired to have a modern airport,[21] and for that purpose organized an architectural competition. The new complex was laid out as a long, single, flat-roofed, three-story building in the style set by the Weissenhof modernists (Pommer, 1983). (Severely austere modern mid-income complexes that were designed by famous architects of the time such as Mies van der Rohe, J. J. Oud, and Le Corbusier.) In its austerity, with standardized windows and structural bays, it became a model of efficiency, but its most exciting feature was the huge visitor deck originated by the architectural gesture. As I have mentioned before, Paris's Le Bourget expansion opened in 1937 and became greatly influential. Nonetheless, the "flat roof" trend was present in earlier examples such as the Schiphol in Amsterdam, the Speke in Liverpool, and the visionary 1929 Leipzig Airport. Hans Wittwer's masterpiece revolved around a futuristic glazed box of stepped viewing platforms, café terraces, and gardens, creating a *soft transition* to the apron (apron is a jargon term for platform). Clearly, this remarkable example was misunderstood in its sophistication and complexity, because in 1936 it became a military facility and was then demolished after the war.

In the *soft transition* between our affair with airports and our present segregation, several elements were transferred from train stations, rail technology, and even sailing. In this regard, Douglas (1995) commented that "the airplane itself was of limited range and capacity, so few imagined that air transport would ever mimic the scale and scope of the railroad. Besides, as the reasoning went, airplanes were different from trains. The image of the airplane was one of freedom of movement in any direction and dimension" (pp. 78–79). At the rise of the "functionalist" architecture, planners, engineers, and designers "urged for the development of airports within existing '*transportation matrices*,' the projects that were being realized provided familiar rather than radical solutions" (Voigt, 1996, p. 13). David Brodherson (1993) has explored at length the formation of the underlying concepts that

[21]Modern in terms of Le Corbusier's idealism, in which planning is "subordinated" to the machine (Boyer, 2003, p. 95).

allowed architects and designers to create the early archetypes before the Second World War. "Initially, designers drew upon forms established in rail transportation before they developed design and planning solutions particular to aviation" (Voigt, 1996, p. 13). This criticism acknowledges that "classicist" architects remained rooted in the past, not just stylistically but technologically. Aviation was a tri-dimensional technology,[22] as opposed to the bi-dimensionality of rail and automobile. Paradoxically, this condition would render impractical the expansion of personal aviation and the dream of a plane on top of every house.

A few modernist architects complained that most airfield terminals of this period "projected the image of a suburban railway station." For example, Glendale terminal in California was nicknamed *Grand Central Terminal*. Voigt (1996) wrote that:

> airplane factories and hangars from this era often took their design cues from the rail-yard shed, and airplane interiors, such as that of the Boeing 80A from 1928–31, imitated that of a Pullman railroad car. The latest aircraft were even put on display in railroad stations, in publicity stunts that were intended to entice more affluent and sophisticated travelers to this new means of transportation, especially as air travel was directly tied to an air-rail transcontinental service provided by Transcontinental Air Transport and the Pennsylvania and Santa Fe railroads. Passengers flew by day and traveled in railroad sleeping cars at night, on cross-country trips that lasted two days and two nights. This speaks of the gestation of early development of a new inter-modal system of transportation, where the airplane covers the long distance and the rail the regional. (p. 13)[23]

The early days of airport planning owe much to the previous experience and success of the rail terminal. Even in the interiors of their airliners, air carriers nonetheless often continued to pay homage to the

[22]Perhaps, this is a topic for further study, but it is certainly fascinating to imagine the complexities and implications of aerial traffic in a Boolean space, versus the straightforwardness of rail or automobile traffic control on Cartesian planes—even in the case of subway systems. By the way, we are still using Cartesian coordinates for air traffic control, and that is why the 9/11 tragedy was technically possible.

[23]Substituting for the transatlantic ship or the transcontinental rail routes.

luxurious sleeping compartments and lounges of streamlined railroad trains until well into the postwar period. In fact, rail industry providers were responsible for equipping aircraft and terminals until the Second World War. Delicate wood finishes on the walls, ample coaches, and overhead nets or bins were installed; even luxurious light fixtures, dining booths, or equipped kitchenettes came from train equipment designers (Delius & Slaski, 2005). Moreover, as I discussed in an earlier paper, "emerging technologies pass through a period of adaptation until they finally become accepted in a more efficient configuration" (Marquez, 2007, p. 17). As written in a *Scientific American* article of the time, "the psychology of the first-class railroad terminal is one of inspiring confidence ... this psychology would be even more desirable at the airport, where every first-time passenger is more or less nervous" (as quoted in Gordon, 2004, p. 43). At LaGuardia Airport, Delano and Licht were sensitive to these issues and imprinted some of their design elements from the architecture of rail.

Because the firm's previous experience was particularly focused on providing comfort to passengers and visitors, they remained faithful to the prevailing notion that the airport terminal was basically a new version of a rail station. In terminals for trains, there is a conceptual/stylistic division between the loading *platform* and the *terminus.* The station building belongs to the realm of the city and its inhabitants; the loading platform belongs to the machine, the trains in this case. Delano intended to follow this same model: His landside architecture is consistent with the nineteenth-century architecture of rail. As the dedication program for the project explains to the opening-day guests: "As Union Station is to the railroads, so is NY Municipal Airport to aviation" (WPA, October 5, 1939a).

Last, in technological terms, Delano proved to have a long-term vision. As shown in his business correspondence of 1939, the architect agreed with the *Office Central des Chemins de Fer Federaux* in Berne, Switzerland, to specifically transfer the baggage technology employed to date in rail stations to his new airport in NYC. An interesting fragment of a letter sent from an executive of the Swiss consortium to its Berne headquarters says:

> Mr. W.A. Delano, cousin du President des Etats Unis, est probablement le plus eminent architecte de l'Amerique. Il vient d'achever l'aerodrome municipal de la ville de New York, et il m'a demande, dessiner l'un ces wagonnets electriques de nos gares suisses, destine au transport des bagages du lieu ou on les enregistre aux trains. (Pasquier, 1939)[24]

Later, Mr. Pasquier drafted orders to adjust the existing electric trolley systems (used at the Swiss rail stations) and sent a proposal to the American architect as soon as possible. The system was successfully incorporated and used initially by TWA and later by other airlines.[25] Surprisingly, this first transfer (adaptation) from rail engineering to airport engineering would trigger one of the most complex systems ever seen in civil construction: "the automatic baggage handling system" of the postwar period.

6 The "Soft Transition"

In this last section, I address the context surrounding architects, engineers, and politicians within the design and construction phases of LaGuardia; I also analyze some specific features of the airport layout that captured for the first time the growing conflict between humans and machines; in other words, structures that "imprinted" (see Latour, 1986) the irrevocable rupture between spectators and airplanes. I briefly describe and comment on the final project and the materialization of the *soft transition*, the implications of the establishment of the landside–airside boundary, and the rise of a rhetoric evidencing the shift toward a new technological culture in airport planning. As I described in the previous section, the WPA division of work became a decisive

[24]Translation: "Mr. W.A. Delano, cousin of the President of the United States, is probably the foremost architect of America. He has achieved the municipal airport of the City of New York, and he has asked me to do a design like the electric wagons of our Swiss rail stations which transport the baggage from one place to another at the register to the trains …".

[25]As observed from a photographic detail of the baggage cart system in a TWA flight. C. Manley DeBevoise, Photographer, Humanities and Social Sciences Library/Irma and Paul Milstein Division of United States History, New York Public Library, 1939, Standard Reference: 1292-D4.

factor in the final configuration of the project. Moreover, the fixed deadline made public by Major LaGuardia added an enormous pressure to the project development. In fact, Colonel Somervell acknowledged that it was literally impossible to start bidding until the final construction drawing set (blueprints) was finished. Nonetheless, the strong relationship between Somervell and Delano grew even closer,[26] thus facilitating the sorting-out of the delicate moment. Starting in 1937, according to a note in the original document set, "approximately one thousand drawings were made in the course of the following eighteen months, together with the accompanying specifications." As I reviewed, all construction drawings were done by hand, drafted on Mylar paper. This was a painfully long process, as we may see in the original drawings (Construction drawing set of New York Municipal Airport and Seaport, 1939), because it demanded crowded draftsman workshops and exhausting reviews to verify the correspondence between the parts and levels. In the office, a summoning by George Licht to his cubicle was famous for "striking terror into the draftsman heart."[27]

Although Delano & Aldrich never considered themselves a large office (maybe compared with the first corporate firms of the late 1940s), the firm's records show that more than three hundred architects passed through its workshop (Pennoyer & Walker, 2003, p. 15). According to the same WPA agency, the full set of documents "included all the complicated drawings for heating and air-conditioning, plumbing, lighting, and inter-communicating systems of telephone and pneumatic tubes. Since it would have required a year to work these out in final form, it was agreed with the WPA & Colonel Somervell, that the drawings should be supplied one after the other and as quickly as the necessary information could be gathered and put on paper. This called for close cooperation between the architects and the WPA Division of Operations" (WPA, 1938, p. 4). I believe the rush to submit the plans could have played a positive role in the innovation, as short

[26]Evident in the active correspondence between W. Delano and Fred Delano from 1937 to 1941, available in the William Delano Collection, The Archives at Yale University Library.

[27]Interview with a former employee of the firm (quoted in Pennoyer & Walker, 2003, fn. 35).

development projects usually receive less pressure from specialists and consultants.

For the non-specialist, LaGuardia Airport sits in an odd location. It meets the farthest end between Flushing Bay and Bowery Bay in a plot of 550 acres along the Riker's Island Channel. Three runways were aligned toward the water and formed a triangle; they linked to a curved apron made of reinforced concrete. Planes taxiing for takeoff or after landing enjoyed the amplitude of the platform. At the central axis, there was the landside terminal, a three-story Art Deco building with a highly ornamented vestibule. At each side, service and office blocks were arranged symmetrically.

The road system led automobiles through a number of wide parking bays up to a rotunda, where arrivals were at the lower level and departures at the main driveway entrance. This system had never been employed before. On either side of the apron, there were monumental maintenance and storage hangars. Uncharacteristically, the three building blocks were "connected" by a flat-roofed promenade that led passengers and supplies directly to the planes. (For the first time ever!) Its roof served as an observation deck. Facing Bowery's Bay, a second terminal served the seaplanes. This building was similar to the landside terminal, but its arrangement was circular and occupied only two levels (WPA, 1938, pp. 5–15).

The general description of the Office of the Major program reads:

> access to the planes is from a covered platform of concrete construction, 1,500 feet long and 200 feet wide, which runs along the edge of the loading apron tracing a long arc between the two groups of hangars and the administration building in the center. Planes will be brought up to a given gate of the platform and passengers can *embark* or *disembark*[28] only a few feet from the gate. A public address system will announce arrivals and departures of planes.

[28]Terms imported from operations at maritime ports. Date from the 1582 French terms *disembarquer* and *embarquer* meaning to remove to shore from a ship, and load onto a ship (Merriam-Webster, 2006).

> Stairways lead down from the concourse on the main floor to this platform. It is from this upper promenade that those seeing travelers off or waiting for incoming travelers will watch the field. The promenade is a point of vantage, open to the public, from which it is expected thousands of persons will view the varied activity of a great airport, with a perfect view of the field at close range and yet safely out of the way of the planes. Five thousand persons can be accommodated on this platform. (WPA, 1938, p. 10)

It is remarkable that, above any other feature of the airport, the infrastructure for visitors was emphasized. I interpret this gesture as the product of LaGuardia's demagogical spirit and sensitivity to pleasing the masses, especially in the rumble of social discontent after the Depression (Kennedy, 2005, Ch. 8). Prior to New York's LaGuardia no airport ever split user *flows* into separate levels, although there are some traceable historical references in the 1928–1929 entries in the Lehigh and R.I.B.A (Royal Institute of British Architects) airport design competitions. Some boards "had included practical recommendations for more subtle systems of boarding passengers, arrangements that were supposed to untangle the confusion on the apron by creating two levels: by means of passenger tunnels that led directly to the waiting airplanes, or by means of bridges from which one descended to airplanes waiting in parallel formation" (Voigt, 1996, p. 45). But it is hard to assign authorship in this matter.

The functional and technical consequences of splitting levels apart were totally unprecedented (see Fig. 1). The firm needed a good aeronautical consultant. Delano hired L. L. Odell as the project's *aviation consultant* (as Odell used to call himself). Captain Odell was an ex-pilot of the golden era, and Juan Trippe's airport planner. He was also Pan Am's Chief Airport Engineer and a key member of the organization; he was responsible for planning most of the company's previous terminals, including the challenging remote projects at Midway, Wake, and Guam Islands. In a letter directed to Delano in October 1939, Odell subtly apologizes for a "little personal discussion," but later changes his tone and shows gratitude for Delano's public recognition of the "value of his contributions" to the project and accepts that as a "reward."

He also feels proud of his technical achievement and claims it will offer the most "efficient operation" of all the airports in the world. Odell changes tack again and acknowledges Delano's "genius" for making the airport "a thing of beauty." He also praises the architect's ability to sort out the "practical and political compromises in happy fashion" in such a "short time that it sounds miraculous" (quoted in Urwin, 2002, p. 27). The letter has a slight undertone of technical condolence, as if the author of the aesthetic were not the author of the planning. Anyhow, it may be inferred that Odell was a very experienced specialist and that he may have contributed significantly to establishing the project layout, and more specifically to proposing the two major innovations of the design. In particular, if we think of his recurrent participation in competitions and projects with innovative ideas, we may assume that he played a key role in this story.

In 1936, the designers and the specialists decided to split flows into two different levels. The imperatives from the WPA (1938) were clear: to "expedite the arrival and departure of domestic airline passengers as much as possible." Dividing incoming passengers at the apron and outgoing passengers into a higher level was the most reasonable thing to do. The logic behind this decision had to do with the increasing complexity of airport systems. In the days of airfields, only a few planes were being serviced or uploading or downloading passengers, so that the implications at buildings were minimal. Restaurants were able to provide food from downtown directly to the airplanes, and city hotels used to send baggage directly to the aircraft compartment, so that the only complexity, if any, was the large number of persons and vehicles visiting the facility. In 1936, LaGuardia declared that "more people reached NYC by air than by sea" (Work Projects Administration, 1938, p. 1), so clearly the influx was already high. Aviation was a growing industry, and the difficulties of providing good service were growing exponentially; hence, the relationship between rail stations and air terminals became a priority. How could they handle hundreds of bags at the same time? How could they serve good meals to dozens of passengers or make them wait comfortably?

The idea of separating "functions" was characteristic of the ideas and writings of the influential Le Corbusier. The Swiss-French planner, architect, and artist stated that the ideal form to re-establish order in the city was to divide its functions and let humans flow among these mega "blocks" of housing, government, education, retail, and so on by employing the latest transport technologies such as the airplane, the automobile, and the train. In between, there would be parks and monumental plazas. In this spirit, human actions were subordinated to the efficiency of the technological system.[29] In the case of New York Municipal Airport, the separation of flows was framed by the potential conflict of synchronizing dozens of operations simultaneously (WPA, October 5, 1939a). The split would reduce human movements as fixed and predictable, in a Cartesian board composed of multiple vectors. The task of the designer was reduced to avoiding conflicts.

Delano, Licht, and Odell decided to organize the terminal in this fashion; they assumed that because airports were independent of the city, most passengers would arrive by either automobile or bus. In this sense, the road system that privileged the automobile and incoming vehicles was ramped up at the outer side of the rotunda; departing vehicles flowed around the inner lane at ground level, so that no obstructions were possible. Therefore, the lower plan of the building held the baggage room, which curiously included a customs section for the first time. On the upper floor, the main vestibule distributed passengers to the waiting areas and airline counters. The challenge was how to bring passengers down to the apron and then make them circulate in an orderly fashion toward their plane. In the case of Le Bourget, the terminal building was an incredibly long building for the time. But there was no chance of following such a scheme, as the square footage available in NYC was much less than those plots available for European airports. The CAA (Civil Aeronautics Authority) and the WPA agreed on the need for docking twenty-one aircraft simultaneously. Therefore, Delano, Licht, and Odell opted to design a *loading platform* "where aircraft parked on the apron and it extended approximately 750 feet at ground level from the center line of the terminal. At the outset this

[29]Le Corbusier's scheme for a new city distribution is presented as early as 1922 in his *Ville Contemporaine*, but does not fully mature until the publication of the *Ville Radieuse* (1935).

enabled airplanes to load simultaneously from a partially enclosed area" (Brodherson, 1996, pp. 77–78), as specified in the original drawing set. "From the air this long curving structure resembled the feeler of a giant insect" (Gordon, 2004, p. 114) that would distribute departing passengers to the airplanes toward the terminal. "In general, enplaning passengers entered the second story of the terminal from the upper-level roadway through the terminal, and proceeded downstairs into the ground-level loading platforms. Deplaning passengers could board surface transportation after walking back to the terminal through the lower level, and out to waiting automobiles and taxis. Alternatively, passengers could walk a short distance across the loading platform from the airplane to waiting automobiles" (Brodherson, 1996, p. 78).

The promenade was rapidly christened the Skywalk, and according to Gordon (2004) consisted of a "long boarding dock that extended from either side of the terminal[;] it was a revolutionary move in airport planning, a precursor to the *pier* extensions that would sprout from terminals after the war" (p. 114). It is important to note that it was not in the original spirit of the Skywalk to divide the apron and the landside activities of the complex. This happened after the opening of the building, when the south side of the Skywalk was partially glazed. It was the product of increasing complexity in the cargo-loading operations. The logic behind this arrangement is that contrary to the rail station, where trains can be loaded and unloaded practically inside a building (or under a roof), airport terminals are severely constrained due to the necessary movements for parking planes. So with the exception of Tempelhof, it was simply too expensive to park aircraft indoors or at least covered under partial roofs or canopies. Thus, when the demand increased, piers became necessary. The Skywalk established a new, two-level relationship with the terminal. In a bold move, Delano & Aldrich decided to split not just the pier but the entire complex. At the terminal, the building's "two levels provided efficient circulation and helped to decrease bottlenecks in the flow. Incoming passengers did not need to pass through the terminal at all but could proceed directly through their gate and out to the taxi stand or parking lot" (Gordon, 2004, p. 114). Essentially, the two-level scheme permitted a functional split between check-in activities (upper level) and the baggage claim (lower level); the

Fig. 3 A view from the Skywalk platform looking toward the apron. Notice the conceptual split between the platform and the parking. Wurts Bros. New York, NY (1944) (Image provided by the Museum of the City of New York)

innovative layout allowed for the very first time the physical separation of incoming and outgoing passengers and systems. I think that the two-level split also simplified the compromised available space for an external drop-off area; in a way, it avoided traffic congestion.[30] Certainly, the architects had in mind new ways of expediting flows: Be it automobiles or buses, passengers, service vehicles and aircraft, and in doing so, they avoided traffic jams, passenger congestion, and thus an overcrowded ramp. It was a good justification for the project indeed.

[30]From analyzing the original blueprints prepared by Delano & Aldrich. Construction drawing set of New York Municipal Airport and Seaport (1939), Avery Drawings and Archives, Avery Library, Columbia University.

As I have presented so far, two new elements were introduced here: One has to do with functional configuration and was the separation of levels, between arriving and departing passengers. The second was an innovative two-level promenade, which allowed passengers to free-flow on the apron and separated spectator flow on top (see Fig. 3). In contrast to the fact that the functional two-level split had purely practical reasons and was mainly a reaction to technical demands, "the Skywalk was not conceived merely for functional purposes. Its curving promenade helped to make the airport feel more like a public place and not an exclusive club for the rich—an important point for a mayor who championed the rights of the common man. It was open to everyone. Turnstiles were installed, and visitors paid a dime to stand on the observation deck and watch planes arrive" (Gordon, 2004, p. 114). (The organizational split was the product of planning more efficiently. The Skywalk bowed more to sociopolitical reasons than to technical ones.)

New York Municipal Airport became one of the largest *modern airports* of its time; its unique "split" configuration responded to the new requirements of a new network of systems. Those were the postal service, aircraft maintenance, baggage, food, ticketing, fuel, lighting, traffic, automobiles, radio communications, electricity, HVAC, and customs control. LaGuardia Airport represents a sort of paradox. On the one hand, it was a small world of systems in itself, gaining in interconnectivity and in those terms an important example of technological change; but on the other it also encapsulated the end of an era. In the New York airport, there is no violent rupture (as envisioned by many, such as Le Corbusier) but rather a *soft transition*. The observation promenade was an extraordinary object that successfully realized the negotiations between two opposing worlds. However, the days of the unbound dream of aviation were fading out, as Antoine de Saint-Exupéry captured precisely in 1939: "progress in aviation technology would inevitably undermine and destroy the moral universe of *La Ligne*,[31] once flying is no longer dangerous, the soul of *La Ligne* will be dead" (Wohl, 2005, p. 201).

[31] *La Ligne* refers to the myth of the intrepid French postal service or *Aeropostale* in the 1920s.

The monumental structure was a practical solution that welcomed the old *spectators* and without having them set foot onto the *airfield* silently transformed them into *visitors*. People were *softly detached* from the physical proximity of the aircraft (see Fig. 4). The spectacular brand-new machines, such as the Boeing Stratoliners and the Douglas DC-4s, stood proudly and perfectly aligned, almost arrogantly observing us. The romantic borderline was vanishing, too; enthusiasts were now separated in spatial planes. The "*air-mindedness*" was clearly at stake, and the airport promenade became a first warning. But passengers who enjoyed an era of fancy courtship will remember fondly those first years of LaGuardia's Terminal. Nurses transformed into pretty stewardesses

Fig. 4 LaGuardia pioneered in separating the different flows. This view shows the courtyard formed between the Terminal and the Skywalk promenade. Notice some ground support vehicles at the forefront, passengers circulating underneath, and visitors walking on top of the structure (Photography by The Port of New York Authority (1955). Donated to the Museum of the City of New York)

outdid themselves to please customers[32]; travelers received gifts, flowers, club memberships, and endless services. As Arend (1979) chronicles, "the airport observation deck was a huge success. Before opening, the city had practically begged concessionaires to operate the deck offering a lease at $12,000 per year, but no one was interested. Later the Mayor had a 10 cent turnstile installed and receipts for the first year came to a tidy $150,000 in dimes" (p. 23). In other words, at least one million, five hundred thousand visitors came to see the Skywalk.

During the opening days, people celebrated and felt welcomed, but the overcast sky and the end of our *affair* were inevitable. (Airport growth in fact put in charge a new generation of engineers and politicians who focused on efficiency rather than aesthetics.) In hindsight, the promenade is almost nostalgic and evocative, even in a photograph. But the Skywalk was not an afterthought. William Delano was not a modernist, and therein lies the uniqueness of this story. He sorted out the pressures of a growing tendency for an austere, social, technology-driven architecture, and designed the terminal much on his own terms, hybridizing styles but without finding much artistic value. In fact, Delano was declared the last classicist architect, receiving the 1953 Gold Medal from the American Institute of Architects, and since that year, no one has received the same honor. It is not hard to imagine the fate of NY Municipal Airport in the hands of the radical modernists of the late 1930s.

Even for Delano, the pressures to design in a more *modern* fashion were strong enough to make him opt for a hybrid style between Art Deco (concourse massing and ornamental detailing), Neoclassicism (concourse vestibule and its reference to the Pantheon in Rome), and Bauhaus-style austerity (repetitive fenestrations, exposed face brick, standard structural bays, visible economy). The building received mixed reviews after its opening. The harshest critique came from the influential philosopher of technology and architecture critic for *The New Yorker*, Lewis Mumford. In a December 30, 1939 *Skyline* column,

[32]See "Perfect Attendants," an excellent historical chronicle of the role of women in air hospitality, by Molly Simms, in the October/November 2007 issue of *Bust*.

Mumford viewed the main terminal as "a series of bungles, missed opportunities and hideous misapplication of ornament [that came] very close to equaling the all-time architectural low." At least, he concluded, "you could get into a plane and fly away" (pp. 49, 50). Also regarding its aesthetic impressions, author Alastair Gordon (2004) has described how an "early rendering by Delano and Aldrich shows a fairly utilitarian façade broken up by horizontal bands of windows and darker brick, but the final version was much more ponderous. Strident vertical ribs and decorative grillwork gave it a vaguely totalitarian look; an impression that was reinforced by the stainless steel eagle perched ominously on the roof. Exterior walls were buff-colored brick, and the coping was black masonry" (p. 112).

In addition, in his *New Yorker* column, Mumford "accused the architects of designing a 'faked entrance' that had nothing to do with the modern split of aviation." Interiors, according to Mumford, were designed in what he described as a "funeral- parlor motif," but he reserved particular scorn for the control tower, which he thought of as a "miniature lighthouse, which had been seemingly plopped on the terminal's roof as an afterthought" (Mumford, 1939, p. 49). On his critique, I think Mumford is basically right, but strictly from an architectural point of view. Several architects of the time expressed similar feelings about the building; however, Mumford avoids technical language and clearly underestimates the less visible achievements of the building. This idiomatic gap was constantly growing wider. The famous critic and philosopher was clearly not aware of the technical pressures and limitations; perhaps he expected an air terminal as grandiose as a rail station. It is a pity that Mumford did not use his recently published "polytechnic versus monotechnic 'framing' to discuss the airport."[33] His known aversion to transportation means such as the automobile would have been seriously challenged by Delano's noteworthy scheme, and particularly by his Skywalk, where evidently not *all* is subordinated to

[33]Mumford (1934) discusses how monotechnics (planes in this case) will only care about its own "trajectory," leaving aside humans and society as a whole. This assumption might be correct in many contemporary airports but was hardly the case in LaGuardia.

machines. In this sense, some of the less visible aspects of LaGuardia are those related to the systematic construction of a new technological rhetoric, on both the technical and official levels. After the completion of the project, a limited set of documents from the WPA was made available. Not surprisingly, it shows an official lexicon that speaks of the growing pressures inflicted from the *airside towards the landside*. Besides the terminology on which I have commented before (airfields, field side, landplane, landside, seaplane, seaside, terminal), other key examples of this rhetoric are worth highlighting. Among the most meaningful expressions relevant to this study, the *Skywalk* encapsulates both the technological demagogy and the real nostalgia for the dream of no boundaries; also, the frequent use of *spectators* instead of the most current *visitors* speaks of a vanishing time (WPA, July 1, 1939b, pp. 4–22). I have found three basic levels of discourse spinning around the creation of LaGuardia: first, the politics of technology and science; second, the rhetoric of technological determinism and technology transfer; and third, the use of language to describe a system or an *organism*.

After studying the original documents that were produced by the WPA for the purpose of diffusion and inauguration of the project, I found that the labels that were used to describe the different rooms and areas of the airport expressed many aspects of the process that are otherwise hard to identify. For example, at the first level of the building, technologies are cleverly disguised as neutral or a-political, but when analyzed deeply in context they seem to reveal a strong political undertone. This finding is similar to Winners's (1986) discussion on the Long Island "low" bridges (p. 20). In the case of the Skywalk, the use of language helps to reinforce the implicit politics of the artifact. Thus, several terms and phrases published in the dedication program for the opening ceremony reveal the political connotations on the technological vision of its time. Here are some of these fragments: "no man can tell the part air transportation will play in the destiny of the new world. No man but knows how important it must be" (WPA, 1939b, p. 10), "the Herculean task was shouldered by Fiorello LaGuardia himself" (p. 10), "the most modern airport" (p. 10), "a control tower with every modern facility of aeronautical research and science installed within" (p. 12), "it stands as a monument to the progress of air transportation and aeronautical research" (p. 12), "it will fulfill a place in American aviation long needed and played its part in helping to further advance this most modern of man's means of conquering time and space" (p. 18). Interestingly, the political discourse is heavily supported by a deep faith in science, technology, and engineering as means to provide a better future.

At the building's second level, we can find the usage of some new terms that may help us to understand how *aircraft systems* gain terrain and displace the previous rhetoric of the rail station or seaports. Terms such as *speeding-up* (processes) often address many *functions* such as maintenance of aircraft but were also employed to speak of passenger flows; the use of *concourse* was imported from rail stations to describe passenger halls. Other words such as *process* or *segregation* appear for the first time, naming the routines of loading and unloading of passengers. Phrases such as "*the very fact that the building has been built for the sole purpose of expediting the arrival and departure of domestic airline passengers*" make clear the new rules of the game.

Others such as "*seaplane hauled on railroad tracks*" did not coin any new terms but became rather peculiar expressions. The maritime world offered references such as *public observation decks* (landside–airport boundary), *docks, piers,* and *admirals clubs* (soon VIP lounges); however, the previous *embark* and *disembark* would gradually go out of fashion; on the landside, the use of the Victorian *driveway* was rapidly discontinued and was replaced with drop-off sidewalk or the introduction of the *baggage room*, which preceded the future baggage-handling zones and supports the fact that a new technological discourse was emerging (WPA, July 1, 1939b, pp. 4–18).

Last, on a third level, language is used to describe metaphorically the idea of an organism or a system and emphasizes the concept of *control* of those systems. For example, the use of phrases such as the *airport nerve center* alludes to a living organism that is controlled from a master brain, or the more explicit *control desk*, which is equipped with a "miniature model" of the whole complex, or other perhaps more technical terms such as "*automatic sprinkler system,*" *baggage system,* "*modern wiring system,*" "*the network of lights*" or "*astronomical time switches*" that appear as language innovations that reinforce the futuristic discourse of the time (WPA, October 5, 1939a, pp. 2, 10, 16). Also, in an interesting letter addressed to Somervell, Delano (1939) shows his growing concern about the recently installed "*automatic sprinkler system*" and emphasizes the previous exhaustive testing. However, the innovative fire-extinguishing device failed a few weeks after the opening of the building.

The interpretation of the emergence of a new speech suggests that language gradually becomes an expression of the forces in conflict. The lexicon progressively becomes more specialized and therefore interlocked, or inaccessible to the non-expert. The expressions mentioned above are only a sample from a wide array of new expressions and neologisms often borrowed from the architecture or engineering of transportation. However, perhaps the most interesting aspect can be found in the first intents to organize human randomness and the strategies to constrain it into a machine-like system that *processes* passengers and employees in choreographed routines. In this sense, language helps to support my initial assumptions. The growing complexity of aircraft, the systematization of airports, the establishment of federal regulations in aviation and airport planning, as well as the business pressures to be more efficient and therefore more profitable, have opened up a stream of progress in which the old romanticism has no place.

7 The End of an Airport's Epoch

Few designs in the history of American technology have been so costly, so successful though so spurious as the Municipal Airport of New York. Its statistics are spectacular: It was the largest airport project ever funded by the State; from its conception in 1937, it went way over budget but was finished on time. After its opening in 1939, the airport was instantly acknowledged and praised technologically, as one of the most advanced airports in the world; it also succeeded socially, as LaGuardia himself advertised it as an airport *open-to-all* for enjoyment and entertainment; and the Skywalk patented that assertion. But, after just one year the complex was working at full capacity, so in 1941 plans for a new airport began.

There is then a strong contradiction between the rapid technological obsolescence of a technological artifact (the airport in this case) and the warm response of its audience (the public and passengers); one would expect the opposite: If a technology is welcome, it will last longer. But this does not necessarily make a rule of thumb. Even though members of social groups may show a high "rate of adoption" (Rogers, 1962, p. 22), technologies could fail because they were thought of as closed systems that cannot be updated or expanded. Nevertheless, LaGuardia Airport can be considered a monumental techno-political device and one of the biggest triumphs of the *American Sublime* (Nye, 1994).

I think it fair to say that the ingenious arrangement of the terminal layout was a vivid example of an intense transitory moment where aircraft and airport systems started to put pressure on the landside–airside divide. As I have shown, politicians, designers, and engineers agreed to the negotiation of a subtle, well-balanced solution, carefully avoiding preference for the emerging systems over people's customs, and in that sense, it became a soft transition and not a drastic one, as many visionaries forecasted. These actions had a clear political value that was capitalized by LaGuardia himself.

As I suggested in this chapter, it is at the Skywalk that two historic cultures clash: the openness of the airfield versus the restrictive nature of the modern airport. This collision remained imprinted in language expressions that went out of use as well as those that remain valid in our times. I could maintain that LaGuardia Airport is hardly a modern building in its essence, but its gestures and unprecedented features were indeed revolutionary for their time (the separation of departures and arrivals on two levels) and set the basis for the new relationship between users and aircraft.

In Nye's terms (1994), the dialogue between spectators and airplanes represents yet again the temporary celebration of the sublime:

> The technological sublime does not endorse human limitations; rather it manifests a split between those who understand and control machines and those who do not. In Kant's theory of the natural sublime every human being's imagination falters before immensity of the absolutely great. In contrast, a sublime based on mechanical improvements is made possible

> by the superior imagination of an engineer or a technician who creates an object that overwhelms the imagination of the ordinary men. (p. 60)

The sublimation of the airport does not come from the architecture and has little to do with the terminal building.

But without the Skywalk nothing near to sublime could have happened there; it offered the possibility to extend the imagination of the common person into the world of aviation, and only then, the skywalker dreams. The bold and sober invitation read that on October 15, 1939, at 2:30 p.m., the Honorable Fiorello LaGuardia would see his dream come true. The special guests would be theatrically received by the landing of the latest aircraft over the reliable airfield, and the deplaning of invitees and exotic tourists. "There wasn't a cloud in the sky and a gentle, northwesterly breeze blew across Flushing Bay" (Gordon, 2004, pp. 110–111, 116). For the opening day on October 15, 1939, an attendance in excess of 325,000 (see Fig. 5) celebrated LaGuardia's triumph (New York City Landmarks Preservation Commission, 1980, p. 2). "The extravaganza is on the boards twenty-four hours a day and has a cast of thousands, a great parade across the largest sky port of them all" (Gordon, 2004, pp. 110–111, 116).

Just a few days after the opening ceremony, the Board of Estimate elected, with a vote of twelve to one, to rename North Beach Airport to LaGuardia Airport (this decision was later endorsed by Commissioner McKenzie; "Backs LaGuardia as Airport Name," 1939). As I mentioned in the first sections, "years before the 'practical use' for the postal service, military and commercial aviation, the flying man-machine myth was largely seen as public spectacle. Spontaneously, people greatly celebrated takeoffs and landings with gigantic welcomes and parades, and this fact originated a profitable industry of entertainment" (Wohl, 2005, pp. 294–301). But the flying man-machine was about to perform for the last time, during the Second World War.

Shadows of war were threatening from across the ocean, but neither Mayor LaGuardia nor Postmaster General James A. Farley admitted that the airport would ever be anything but a place of peace. "It is fortunate for us here in the United States that we can gather to dedicate an airport of this character to the pursuits of peace without thought

Fig. 5 This plate of LaGuardia Airport pictures the thousands gathered at the opening day. *Dedication of New York Municipal Airport* (1939) (Reproduced with permission of *NY Daily News*/via Getty Images)

of war, of conquest, of the dread possibility of armed conflict" said the Postmaster General (quoted in Arend, 1979, p. 21). He hoped the time would never come when the airport would be needed for military purposes. "This airport is a strong guarantee of the nation's peace and safety," concluded Farley (quoted in Arend, 1979, p. 21). Along the same lines, Washington National was built just one year after its northern cousin, and it represented a refinement of LaGuardia's scheme and its landside–airside boundary.

Nonetheless, the dialogue was finally broken. William Delano and George Licht designed a new scheme for Idlewild Airport in Queens, now known as JFK. The beautiful illustrations by Hugh Ferris speak for themselves; borrowing from the dramatic portico at Tempelhof and perhaps influenced by the monumentality of Albert Speer's masterpiece *Zeppelintribüne*, Delano offered a more refined, operatic version than his Skywalk in LaGuardia. In his plans for Idlewild, the architect delivered a more direct, open contact for the *air-minded* spectator; but at the dusk of the Second World War the grim shadow of national security darkened the skies, and Delano was forced to break his relationship with the Port Authority in 1948. LaGuardia and Washington National became the last true examples of the romantic borderline. As air travel became a risk-free activity, these two airports became congested and thus obsolete. As Saint-Exupéry predicted, without risk "*La Ligne*"[34] was dead, and with it, much of the romanticism of an epoch. As the airport reached the end of its first cycle, the modern landside–airside frontier was finally born.

References

Publications

Arend, G. (1979). *Air world's great airports: La Guardia 1939–1979*. Published in New York, NY, by Air Cargo News. Long Island, NY: LaGuardia Archives at LaGuardia Community College.

Backs LaGuardia as airport name. (1939, October 29). *New York Times*.

Bayor, R. H. (1993). *Fiorello La Guardia: Ethnicity and reform*. Arlington Heights, IL: Harlan Davidson.

Bednarek, J. R. D. (2001). *America's airports: Airfield development, 1918–1947*. College Station, TX: Texas A&M University Press.

Bednarek, J. R. D., & Bednarek, M. H. (2003). *Dreams of flight: General aviation in the United States*. College Station, TX: Texas A&M University Press.

Boyer, M. C. (2003). Aviation and the aerial view: Le Corbusier's spatial transformations in the 1930s and 1940s. *Diacritics, 33*(3–4), 99–116.

Brodherson, D. (1993). *What can't go up can't come down: The history of American airport policy, planning and design* (Doctoral dissertation). Cornell University, Ithaca, NY.

[34] "*La Ligne*" refers to "The Line," or the earliest French airmail system—established through an incredibly dangerous trajectory between Toulouse and Santiago, and which was later known as the Aéropostale (Wohl, 2005, p. 201). The author of "The Little Prince" thought that it was risk, and not necessarily the need to deliver mail, that moved those pilots.

Brodherson, D. (1996). An airport in every city. In J. Zukowsky (Ed.), *Building for air travel: Architecture and design for commercial aviation.* Munich, Germany and New York, NY: Prestel and Chicago, IL: Art Institute of Chicago.

Chadeau, E. (1995). History of airports. In W. M. Leary (Ed.), *Airships to airbus: The history of civil and commercial aviation.* Washington, DC: Smithsonian Institution.

Clack, V. A. (1998). *Airports.* Grove Art Online. City, ST: Oxford University Press.

Construction Drawing Set of New York Municipal Airport and Seaport. (1939). *Avery Drawings and Archives.* Avery Library Columbia University.

Corn, J. J. (1983). *The winged gospel.* New York, NY: Oxford University Press.

Cortissoz, R. (1924). *Portrait of ten country houses by Delano & Aldrich. Drawings by Chester B. Price.* Cornell Library.

Delius, P., & Slaski, J. (2005). *Airline design.* Dusseldorf, Germany and New York, NY: TeNeues.

Douglas, D. (1995). Airports as systems and systems of airports. In W. M. Leary (Ed.), *Airships to airbus: The history of civil and commercial aviation.* Washington, DC: Smithsonian Institution.

Eggebeen, J. (2007). *Airport age: Architecture and modernity in America* (Doctoral dissertation). City University of New York, New York, NY.

Garvey, W., & Fisher, D. (2002). *The age of flight.* Greensboro, NC: Pace.

Gordon, A. (2004). *Naked airport.* New York, NY: Metropolitan Books.

Higham, R. (1995). A matter of utmost urgency. In W. M. Leary (Ed.), *Airships to airbus: The history of civil and commercial aviation.* Washington, DC: Smithsonian Institution.

Hughes, T. P. (1983). *Networks of power: Electrification in western society 1880–1930.* Baltimore, MD and London, UK: Johns Hopkins University Press.

Kennedy, D. M. (2005). *Freedom from fear: The American people in depression and war 1929–1945.* New York, NY and Oxford, UK: Oxford University Press.

Kline, R., & Pinch, T. (1996). Users as agents of technological change: The social construction of the automobile in the rural United States. *Technology and Culture, 37*(4), 763–795.

Latour, B. (1986). *Laboratory Life.* Princeton, NJ: Princeton University Press.

Manners, W. (1976). *Patience and fortitude: Fiorello LaGuardia.* New York, NY and London, UK: Harcourt, Brace & Jovanovich.

Marquez, V. (2007, November). *Why new materials take old shapes: The case of the first iron-bridge in 1779.* Conference paper presented at the 2007 meeting of the Swiss Association for the Study of Science, Zurich.

Mumford, L. (1934). *Technics & civilization*. Chicago, IL: University of Chicago Press.

Mumford, L. (1939, December). The sky line. *The New Yorker*, p. 49.

New York Municipal Airport. (1938). *A detailed description of the great air terminal by the Work Projects Administration* [Original document]. I.T.S. Library, University of California at Berkeley.

Nye, D. (1994). *American technological sublime*. Cambridge, MA: MIT Press.

Pasquier, R. D. (1939). [Letter to the Office Central des Chemins de Fer Federaux in Berne, Switzerland]. *William Delano Collection*. Archives at Yale University Library.

Pennoyer, P., & Walker, A. (2003). *The architecture of Delano & Aldrich*. New York, NY and London, UK: Norton.

Perkins, C. (1907). When we fly: How the accomplishment of aerial navigation will make necessary a revision of human laws and customs. *Haper's*, pp. 2–3.

Pommer, R. (1983). Revising modernist history: The architecture of the 1920s and 1930s. *Art Journal, 43*(2), 158–169.

Proust, M. (1954). *Swann's Way*. New York. Heritage Press.

Rogers, E. M. (1962). *Diffusion of innovations* (4th ed.). New York, NY: Free Press.

Stapleton, D. H. (1987). *The transfers of early industrial technologies to America*. Philadelphia, PA: American Philosophical Society.

Szurovy, G. (2002). *The art of airways*. St. Paul, MN: MBI.

Szurovy, G. (2003). *The American airport*. St. Paul, MN: MBI.

The New York City Landmarks Preservation Commission. (1980). *Report*.

Toulier, B. (2000). *Berlin Tempelhof, Liverpool Speke, Paris Le Bourget: Casisse national des monuments historiques et des sites*. Paris: Patrimoine.

Urwin, G. J. W. (2002). *Facing fearful odds: The siege of Wake Island*. Lincoln, NE: University of Nebraska Press.

van der Linden, F. R. (2002). *Airlines and air mail: The post office and the birth of the commercial aviation industry*. Lexington, KY: University Press of Kentucky.

Vincenti, W. G. (1976). *What engineers know and how they know it: Analytical studies of aeronautical history.* Baltimore, MD and London, UK: Johns Hopkins University Press.

Voigt, W. (1996). From hippodromes to aerodromes. In J. Zukowsky (Ed.), *Building for air travel: Architecture and design for commercial aviation*. Munich, Germany, New York, NY: Prestel and Chicago, IL: Art Institute of Chicago.

Winner, L. (1986). Do artifacts have politics? In L. Winner (Ed.), *The whale and the reactor: A search for limits in an age of high technology*. Chicago, IL: University of Chicago Press.

Wohl, R. (2005). *The spectacle of flight.* New Haven, CT and London, UK: Yale University Press.

Work Projects Administration. (1938). *A detailed description of the great air terminal.* New York Municipal Airport. I.T.S. Library University of California at Berkeley.

Work Projects Administration. (1939a, October 5). Dedication. *Program of events.* New York Municipal Airport. I.T.S. Library University of California at Berkeley.

Work Projects Administration. (1939b, July 1). *Report of activities.* New York Municipal Airport. I.T.S. Library University of California at Berkeley.

3

"Reinventing the Airport?": *Annex 14*, Dulles Airport's "Mobile Lounge" and Other Jet-Age Paradigms, 1946–1962

In our time, it would sound near to impossible that any private or public entity could dare to propose and invest money in reinventing any consolidated system such as the road network, maritime ports, rail stations, or even the subway. Publicity stunts such as Elon Musk's Hyperloop project will probably end up in shelves just like the infamous Aramis rapid transit did in 1987. It is simply too expensive and the risk involved seems to be monumental. However, back in the late 1950s, some countries started thinking in sponsoring a radical new departure from the airport model of those days. This pictured the size of the problem created by congestion. All promoters knew that even with the standardization of the aviation system at stake, and in spite of that risk, they moved forward. This chapter dusts off and analyzes this story. I speak, in seven sections, of two themes that are interlocked between each other: the context of airport planning after the Second World War and, in consequence, the second reinvention of the airport. More in detail, in the first four sections, I frame in general, the problem of planning and designing airports in the postwar period. First, I present the impact of the ideology and the discourse of the postwar period; second, I comment on the formalization of airport planning as a profession and describe some of

V. Marquez, *Landside | Airside*, https://doi.org/10.1007/978-981-13-3362-0_3

the methodological concepts, mainly from S&TS, that I use as tools of analysis. Third, I describe the importance of language in the early days of expert planning, and fourth, I argue that the airports of the 1950s and 1960s gained "momentum" in the optics of Hughes's "large technological systems model." In Sects. 5–7, I develop the case study of Dulles International Airport in Washington, DC, which was planned and built between 1958 and 1962 (see Fig. 1). In Sect. 5, I present the result of my bibliographic and archival research (based on both primary and secondary sources) of the second reinvention of the airport, this time at the request of the US President's Office and through the FAA. In Sect. 6, I discuss the process of reinvention and the birth of the "Mobile Lounge," and last, Sect. 7 addresses the materialization of a new kind of landside–airside boundary at the new Washington airport.

Fig. 1 Artist rendering of Dulles Terminal by Eero Saarinen. This is one of the few modern buildings ever depicted on a US stamp. Solodov, Aleksei (1982). *A stamp printed in the US shows Dulles Terminal* [digital] (Reproduced with permission of Shutterstock)

1 Post-Second World War Airports

As most modern airports became severely congested right after the war, the future of airport planning after the construction of LaGuardia became hard to predict. The objective of this chapter is to understand why the following two decades after 1945 lead the way to the formal (official) reinvention of the airport (Dulles) and how the landside–airside boundary plays a role in this timeline of change. For many historians, airports progressed throughout time thanks to a process of innovations building on newer innovations kind of loop, until high degrees of airport efficiency were thus reached. For architects and engineers, this process lies, in practice, in the construction of an airport typology during the postwar period.[1] However, there are important signs that historical design changes in the airport may have occurred otherwise. The main hypothesis in this section suggests that there are at least three key aspects that "tend to resist changes in the direction of development" (Hughes, 1983, p. 140):

1. The fact that airports gained technological "momentum," creating a socio-technical system that is visible in the consolidation of airport planning as a specialized discipline (this movement is permeated by the discourse of the Cold War).
2. The fact that airport growth triggered initiatives of regulation and normalization in order to stabilize the infrastructure as a system and later as a system of systems.
3. The rising complexity of airport operations and the introduction of new technologies. (New technologies such as jet aircraft, automatic systems, and new building methods and materials brought passenger discomfort, awakening a strong public reaction to noise, disorientation, congestion, extenuating distances for walking, long lines, and exposure to the elements.)

[1]This classification among professionals evaluates and recommends what is working best or is not in the field; it consolidates certain schemes and solutions and agrees on their efficiency; last, younger designers can pick up where the previous expert left off and move ahead, making only minor adjustments. This is the route map toward what I call "standard airport design."

4. In the reformulation of the modern city of the postwar. [Airports were labeled by social and political groups as the new cathedrals of our time or the urban *gateways* of our cities. These metaphors and social constructions will pave henceforth the way to crisis and what I will later refer as a "reverse salient" (Hughes, 1983) in the consolidation of the airport as a large technological system.]

More specifically, I use the airport culture of the post-Second World War period as a framework to understand the growing demands of aircraft innovation (jet propulsion) and the implications of the air travel boom of commercial aviation during the 1950s and 1960s. The airport history author Douglas Karsner (1993) discussed how the period of economic growth after the Second World War altered the industry of urban development and tourism. As the number of airplanes grew, so did the number of passengers, in the same proportion, pushing the terminals to stretch endlessly as long docks. Passenger discomfort became a constant, and saturation and delay took over the airport. In addition, the newer equipment produced considerably more noise, adding to the long list of users' complaints. The pressures on the landside–airside boundary grew steadily, greatly straining the interrelationships within the industry. Designers and experts responded with mechanical gadgets such as the 1952 "Loadair System" by Whiting Co., the 1954 "Air dock" prototype by United Airways, and the installation of the first *Jetway* by JBT Co. in 1959. Meanwhile, politicians and investors became more concerned with the public image of the airport as an icon of progress. Terminals were transformed as the new "urban cathedrals" (Downs, 2009, p. 1), and in this regard, TWA terminal in New York became an inspirational model to cities worldwide.

Paradoxically, the building's sound public success placed its architect, Eero Saarinen, at the forefront of an FAA initiative to develop new ways to deal with the rising landside–airside boundary conflict. How to reduce *delay*? The answer was to rethink the airport as a technical system. This chapter puts particular emphasis on the experimental planning behind the new Washington Dulles Airport in the late 1950s. In this section, I recount the era's faith in technology as a vehicle of progress and, more particularly, how the invention of the "Mobile Lounge" was seen as the ultimate solution for the modern airport. Based on the use of primary and secondary sources, this study goes deep in search of the causes and ideas behind a radical reformulation of the landside–airside boundary. Why push change at such enormous cost?

In the next sections, I use Thomas Hughes's (1983) metaphor of technological momentum to describe how a supportive context or culture led to the development of a super-system of airports "—a socio-technical one" (p. 140), and how interpreting Hughes's model Dulles, just like LaGuardia, may be understood as a "reverse salient."

In a fast-changing industry such as aviation, and particularly after its technological boost during the war years, most of the modern prewar airports in the USA became rapidly obsolete, and "the changing nature of the airplanes that use them created many problems for the airport designer" (President's Airport Commission, 1958, p. 39). The advent of four-engine airliners and the eventual arrival of the German jet engine to commercial aviation marked a new technical challenge for the existing facilities across the country. One may be tempted to think that the pressures of the new equipment, the growing complexity of airport systems, and the new dimensional requirements were ultimately responsible for the new concept of the postwar airport. Authors of planning manuals and technical texts during and after the war seemed to rely on these matters and attributed these tendencies of change solely to the "rapid rate of growth and changing demands of new aircrafts" (Froesch & Prokosch, 1946, p. 20). However, in this investigation, I have found that this is only partially true and that above the forces of technological progress, the larger discourse of the postwar era was perhaps the primary agent of change.

In terms of planning, the demand for new airports came from two main sources. The first is rooted in the National Airport Plan, which in the early 1940s anticipated significant air traffic growth and therefore recommended initially the construction of 470 new airports, although the number of projected new facilities and renovations grew to a huge 3500. Although the plan was overtaken by the Second World War, it was readapted to the new needs of the country. According to airport historian Geza Szurovy (2003), the plan that "ultimately emerged was skewed geographically to military rather than civilian needs," although there were "myriad requirements for converting an airport for military use or building one from scratch, depending on its intended mission"; and the fact that the war "launched the biggest concentrated industrial expansion in America's history, which sent the demand soaring for airports. Domestic and international air transportation and freight operations became critically important to accomplish certain tasks with the urgency required by modern warfare. From requisitioning about 170 airliners at the outbreak of the war, the military air transport system grew in the next four years into an international airline operating thousands of aircraft on a network of routes spanning the globe" (Szurovy, 2003). In addition, although several airports were benefited by its military use—in terms of expansions and renovations, according to historian Janet Bednarek, "the defense-related aid during the 1940s brought significant improvements to airports all over the country. And the continued military value of airports provided part of the justification for the postwar airports aid program" (Bednarek, 2001, p. 11).

The second source of demand for new airports came from the Federal Airport Act of 1946. An extended network of airports would guarantee a national coverage in terms of an eventuality, but who was going to pay for it? In this sense, it was evident that the available public resources were not enough to cover such a massive enterprise, so the federal government had no choice but to promote a municipal plan of financing. For this purpose, two aspects played a significant role: the construction (re-construction?) of a culture of an air age armed with phrases such as the skyways of tomorrow and the utopian dream of an airplane in every home and an airport in every town as flagships. The second aspect came from conditioning development to technological progress: "a community's future position may depend upon its airport planning today. Failure to provide airport facilities may not necessarily mean another ghost town, but it is too big a chance to take" (Johnson, 1946). In contrast, "towns and cities were encouraged to build new airports because they were 'relatively low-cost' and the modest investments would become profitable considering the future revenues from rentals and concessions" (Bednarek, 2001).

Within this context, the first airport-planning manual of the postwar era was austerely named *Airport Planning*. This volume was edited in 1946 by John Wiley & Sons of New York and authored by two employees of Eastern Airlines at the time, Charles Froesch, who was chief engineer, and Walther Prokosch, who was head of the planning unit. Although a number of publications such as the annual *Airport Construction and Operation Reference* preceded *Airport Planning*, this work became a wide-range planning source of the time. The book is a comprehensive treatise on airport planning including sections on planning, technical requirements, and construction, but it also includes intriguing sections in which the authors give themselves license to theorize on the larger historical context of airports. These aspects are especially interesting, because the undertones of cultural history are often more transparent in the technical literature. My analysis of *Airport Planning* reports several findings at different levels.

The first finding is the total omission of the terms landside, airside, and seaside. The terms are not even included in the extensive glossary at the end of the book. In the manual, there is scarce mention of the quintessential LaGuardia, which at the time was clearly a model no longer to be followed. Broader mention is made, however, of the marine terminal, in which the term seaside has been suspiciously substituted for seaplane (Work Projects Administration, 1938, pp. 238–239). I suspect that the authors deliberately rejected the airport divide, and thus the terms I mentioned before. This is consistent if one considers their efforts to understand the airport as something larger and interconnected. But at the same time, they envisioned the airport and its traffic as something subject to be forecast through the use of mathematics and scientific methods; the planning of the airport needed to pass from the hands of artists to the hands of planners and engineers.

Thus, the second finding is the growing need to conceive airports as systems. Throughout the different essays, the words network and system are used in multiple ways. Initially, the authors describe aviation through the concept of airline standard operation, which implies the consolidation and maturity of a system; the emphasis is on stability and gradual change. Mention is also made of patterns in aviation, which in a way implies that the evolution of the technology is somewhat foreseeable and risk-free. With these ideas as a backdrop, the authors present the concept of a national network of transportation and specifically how the airport will provide access to smaller communities and connect them into the net. They also emphasize not considering aviation as an "agent that will relegate existing means into the shadow of obscurity" (Froesch & Prokosch, 1946), but the contrary that aviation will become part of this large-scale system. In terms of planning the continuous expansion and spread of airports and its parts, the other elements of network will link them together: express highways and the automobile, subways, and trains. But perhaps the network is better defined with this metaphor: "a system of arteries and veins is planted in the body of this new creation so that its lifeblood the population may travel freely from one portion to another"; the network is the physical infrastructure that allows the system to operate, and uncertainty and risk are therefore reduced to a minimum.

The third finding is an open concern for making airports useful parts of the system of national defense. For example, when speaking of the future airport planner, Froesch and Prokosch (1946) maintain that, "to achieve the best engineering compromise, he (the planner) must appraise all factors which govern aviation as a means of transportation and must try to weigh what were once imponderables. He must direct his efforts toward planning facilities which will be useful units of an integrated whole, forming a broad pattern of civic usefulness and national defense" (p. 8). Here, the authors clearly acknowledge the dual nature of the airport as a military and civil facility, and, therefore, they extend this idea as core to the new airport planning profession.

The fourth finding is the need to make airport designers aware of urban decentralization and the new policies of dispersal. On several occasions, direct reference is made to decentralization, particularly in three different contexts. First, Froesch and Prokosch (1946) credit the automobile as the major force behind urban decentralization. Second, they use the term dispersal while issuing recommendations for decentralizing airports and transportation in general. The third use of the concept of decentralization is made in reference to the modern concept of the city, where the functions must be dismembered and separated from the core. The authors argue in favor of this split and suggest the construction of at least three airports in every city. One airport must be in the residential suburbs, and its main purpose must be personal travel and recreation. The next is the commercial airport, a facility that should attend to business needs. The last is the cargo airport, which should be placed in relation to the most advantageous location to industry.

The last finding is perhaps Froesch and Prokosch's (1946) most ambitious claim and in a way is the best example of what I call the ideological undertone of the work. The authors make clear throughout the different sections that the real challenge in airport planning is to predict future changing conditions and to avoid obsolescence at any cost. Therefore, they introduce the concept of a true "air city," a larger metaphor that perhaps unintentionally captures many of the dreams and fears of the epoch. The "air city" will supersede the prevailing "motorcar city": "The disfigurements which the motorcar, plus lack of planning, has visited upon the American scene are well known. The beauties of landscape which are opened up for us by the motorcar and the opportunities for fuller living are equally well known" (Froesch & Prokosch, 1946). The authors maintain that the airplane will affect our "individual and national lives even more rapidly than the motor vehicle" (Froesch & Prokosch, 1946). The airport city is a city where planning is subordinated to the airport. It is developed on a vast space, dividing the functions of the city—in a way very similar to the proposed scheme by Norbert Wiener published in *Life* magazine in the early 1950s, in which the city parts are separated and linked through highways or "life belts" (Boyer, 1985); the main difference is that contrary to Wiener, the airport becomes the center of distribution (passengers, products, and objects).

The ideology of the early postwar era was ever present in the conception, or re-conception, of the modern airport. In addition to the technological advancements—many of them products of the runaway warfare industry during the war—I suggest that the machinery of progress was not solely, or even mostly, responsible for the construction of a new airport culture in the late 1940s. In presenting evidence from technical sources, I suggest that within the technical discourse, much of the period's political ideology remain embodied in these documents. But perhaps what best explains why it was necessary to diffuse this ideology is the fact that the proliferation of airports was linked to the plans of urban spread and, at the same time, the necessity to involve the municipal governments in order to finance the initiative.

This was clearly not just a technical move; the underlying reasons were well founded in the incapacity of public funds to cover the huge costs of the expansion. What is evident to me is that without the construction of metaphors and the prediction of future views that made airports simply necessary, the government would have not made this rapid expansion possible.

2 Formalizing Airport Planning

Just a few years later, at the turn of the 1950s, airport planners claimed that forecasting and regulating the airport were indispensable to avoiding the alarmingly rapid obsolescence of airport facilities. It is timely to remember that when compared to the rail station, where even the newer trains adapted to the old tracks and station sheds, the newer airport inevitably ends in need of having large investments put to no use, in just a few years. In 1947, the influential *Architectural Record* published the article "Establishing an Airport Planning Program," in which Smith, Hinchman & Grylls, Inc., affirmed that "the tendency in the past has been to lay principal stress on the engineering factors of airport planning with no program established to guide the engineer and architect in the long-range requirements of an airport. Previously built airports have been used as guides to space-use patterns. These examples no longer serve that purpose." Thus, the lack of a set of airport planning principles or a theoretical framework somehow pushed airport designers to base new ideas on previous cases in a sort of "learn by using" (Rosenberg, 1994) or learn by error model; thus this continuum of real-world constraints and use of technical logic shaped those "technological trajectories" (Vincenti, 1976) so often found in engineering thinking. The airport was thought of, then, as a scientific issue or a problem with a predictable input and output.

As I will discuss later in my analysis of Dulles, planners and designers reached the extreme of documenting cases on five continents in order to close all possible windows to potential failure. My contention in this regard is that this method masked the timely identification of the landside–airside boundary as a technological anchor.

The influential Walter Prokosch—the co-author of the first planning manual that I mentioned before—wrote in a long essay in 1951 about the importance of programming through experts: "The evolution of a program for the various facilities needed on an airport consists of a complex series of calculations which are generally prepared by the airport specialist" (p. 113). It is interesting that he speaks of the evolution of a program as an organic, ever-changing entity. True, it is in that process of refinement and readjustment of the airport/terminal programming that the core of this thesis argument lies. What Prokosch calls evolution is close to what I define in this thesis as a negotiation process. However, in practice, this is not that simple, and the so-called programming[2] falls often into several hands at the same time.

But planners were busier figuring out methods to foretell traffic in order to increase financial "certainty" than understanding the growing landside–airside conflict as a socio-technical construct. In 1953, *Progressive Architecture* published a special issue, "Airport Terminal Buildings." In it, the editorial staff said that "after community needs, air traffic, and economic considerations have been studied, the program planners will finally come to the great question: How large should the terminal be? The key to the answer and all its parts (how many aircraft loading stations; how much passenger waiting space; how much space for toilets, restaurants, concessions; how much additional space for visitors and sightseers) lies in an estimate of peak-hour traffic, according to presently accepted planning methods" (Arroyo et al., 1953, p. 70). But as in all rule-of-thumb calculations, "various unusual factors will affect the final figures, and the method of calculations are not all agreed upon" (Arroyo et al., 1953, p. 70).

Technically speaking, peak-hour passenger traffic is translated into aircraft space requirements, which inform the spatial program. Peak-hour passenger traffic is divided by average aircraft seating capacity (a variable, depending on type of airport according to CAA during the 1940s and 1950s); the result represents the number of peak-hour

[2]In practice, the airport terminal programming of spatial, technical, human requirements is usually subordinated to a preliminary cost analysis.

runway operations. Therefore, the building requirements can be almost directly translated from the peak-hour passenger figure. In the same issue, *Progressive Architecture* made reference to this aspect: "although passengers and visitors mingle in some space, the ticket counters and the baggage pick-up space are related directly to the traffic count" (Arroyo et al., 1953, p. 70). The fact that planning became primarily an analytical activity left aside the understanding of multiple dimensions of the landside–airside boundary, and thus, it was put under an enormous strain.

According to the experts of *Progressive Architecture*, the architect may adopt either of two options: "The smaller port, with its lower passenger volume, may use a frontal arrangement with aircraft strung out along the perimeter of the building" or what I call here linear arrangements or parallel piers. "For the larger-volume terminal, a finger scheme, consisting of piers stretching onto the field, perpendicular or radial to the building, with planes grouped along these extensions, seems best." Among the benefits of this scheme was the chance to have a centralized ticketing and baggage-handling operation for independent airlines (per pier), "since walking time from the waiting room and loading distances from baggage and cargo rooms can be reduced" (Arroyo et al., 1953, p. 90). These kind of geometrical arrangements were desperate measures to counterattack delay and the long walks.[3]

Perhaps one of the major impossibilities for studying the landside–airside boundary as a design problem was the fact that too many actors used to intervene in decision tables. By the 1950s, *Progressive Architecture* experts discussed the reasons for which designers felt they were not getting the job done: "In business practice, as in planning and design, the airport commission is likely to be a complicated one for the architect, as well as for the consulting structural, mechanical, and electrical engineers who works with him. In crafting the first-of-a-kind monograph, the magazine's editors have talked to many experienced architects and engineers in this building typology, and opinion was unanimous: the multiplicity

[3]Walking distances were already causing sound complaints from passengers, among the most sound were the ones against Chicago O'Hare that used to have pedestrian trajectories of over one mile.

of clients to be satisfied makes approvals difficult and changes likely; relationships with consultants hired directly by the airport authority as experts on one phase or another of the problem are not always pleasant; fees, generally, are too low" (Arroyo et al., 1953, p. 113).

Among the multiple clients or groups, agencies, and individuals pushing requirements that should be fulfilled are the actual client (municipal and federal governments), representatives of the different scales of government such as state, county, and municipal; the airlines, the CAA, US Post Office, US Weather Bureau, US Public Health Service, Bureaus of Customs, Immigration, DIR, Department of Agriculture, local fire and police departments, a myriad of specialists, and the main concessionaries.[4] The required negotiations between the numerous actors were characterized then by the existence of pressures and slowness ("The airport scramble," 1956, p. 133).

Planners started studying the moving patterns of humans and called them flows. These passengers' circulations were the primary basis for planning.[5] During the 1940s among the studied flow patterns, the need for a physical separation between airport visitors and passengers arose as perhaps the most critical and unpopular ("Activities are channelized and separated," 1941, p. 53).[6] In describing the future airport of the 1950s and 1960s, the firm Smith, Hinchman & Grylls, Inc. (1947) explains the desired sequence: "The enplaning passenger leaves his car or airline limousine at road level and point of baggage check-in and proceeds by escalator to second level ticketing, lounges, rest rooms, or directly to gate of plane departure" (p. 93).

At this point, a passenger is checked in by a ticket agent and boards his plane by way of an all-weather gangway. Guests of air passengers have access to an air passenger lounge. "The public has a separate entrance and a spectators' promenade at second level in common with shopping and recreational facilities" (Smith, Hinchman & Grylls, Inc.,

---

[4]See who gets involved in the subsection "The Client" (Arroyo et al., 1953, p. 116).

[5]See the full flow sequencing on page 116 of Arroyo et al. (1953).

[6]The literature also refers to the term "channelized," to describe the conduction of many activities, but particularly baggage.

Architects and Engineers, 1947, p. 93). Logically, there were strong consequences in the landside–airside boundary dimension, being perhaps the more evident when planners assumed that aircraft were what really interested visitors and not the social spectacle of traveling (Arroyo et al., 1953, p. 129). The spectators' fascination with the theatricality of travel (having the airport as a backdrop) is evident in such memorable stills as those taken when the Beatles arrived in America in 1964.

During the early postwar period, planners coined new terms that became part of the technical jargon and help to build momentum. *Progressive Architecture*'s influential special issue on airports in 1953 described how the planner must insure that international travelers are "processing through customs, immigration, and other agencies that are concerned with their movements or the contents of their baggage. If they are incoming, they must be segregated, and their baggage must be handled separately" (Smith, Hinchman & Grylls, Inc., Architects and Engineers, 1947, p. 94). I highlight the use of *processing* and *segregation* as an early usage of the terms.

One more aspect of this special issue that speaks of the struggles between economic and social interests is when the authors speak of "the balance between necessary income and passenger convenience must again be carefully weighed" (Smith, Hinchman & Grylls, Inc., Architects and Engineers, 1947, p. 92). This idea addresses how decision making will often subordinate users' comfort to economic interests. During the mid-1950s the CAA (Civil Aeronautics Authority, later FAA) issued recommendations that called for "economic practicability, amortization potentialities, and functional efficiency should be the prime considerations in the allocation of space. Expansibility and flexibility should be the prime considerations of design and construction" (U.S. Civil Aeronautics Administration, 1953). After the bitter experiences of the previous years, it was clear that the state machinery aimed for healthy finances and expandable facilities. But this speaks specifically of the interests of CAA, who played a double role as the responsible governmental agency and principal client at the negotiation table. But as I show later on, this does not reflect the interests of the US government as a whole. Other politicians had other ideas in mind.

Perhaps located in what critic Lewis Mumford (1934) called "the fantasies of the future which have been suggested by the triumph of the machine" (p. 492) is the intriguingly common ideal of thinking of the airport "as basically a machine to facilitate transfer of passengers from the ground to the air and vice versa" (Arroyo et al., 1953, p. 88), as described by the *Progressive Architecture* editors. The comprehensive study in the American magazine assures that the "ideal airport terminal would be a smoothly functioning machine. But what should flow with machine precision, too often has become over elaborate and complex" (Arroyo et al., 1953, p. 82). The publication ends making no mention whatsoever of the terms landside and airside, although this absence seems to be consistent with many of the articles and manuals of the epoch.

3 The Language of Planning

So far I have described how during the war and at least until the mid-1950s the terms landside and airside were either put into disuse or deliberately abandoned for not reflecting the new interests of airport planners and their mechanistic vision of the airport. The Oxford English Dictionary credits the *Daily Telegraph* in 1955 as the first publication to employ the words landside and airside in order to define the distinct parts of an airport terminal, but as I have shown in chapter one, they were used during planning stages of LaGuardia Airport at least eighteen years earlier, in 1937. Proving that one exception always breaks the rule, the normative bible of airport planning and design, the famous *Annex 14* of 1947, surprisingly makes constant reference to both terms. Thus, in these following paragraphs, I will elaborate on how and why the terms had such a preponderant role in the manual.

The imperative need to regulate the chaotic disparity of national and international aviation encouraged the international community to found the 1944 PICAO (Provisional International Civil Aviation Organization), which, after two years of analyzing the status of international aviation and airports, took its final and current shape as ICAO (International Civil Aviation Organization) in 1946. Its standards for *aerodromes* (notice the word, instead of airports) were instantly accepted worldwide. The same team of international experts put on paper a number of Annexes or standard manuals that were meant to regulate both the airspace and the land-space of aviation. The first edition of *Annex 14, Aerodromes* (1951), was rapidly replaced by the thoroughly revised second edition of 1953—which, according to contemporary airport planners, instantly became the norm for planning and designing airports (see next chapter). The terms landside and airside are used throughout the 133 pages of the document and even defined in the separate glossary of terms:

> Airside: "That area of an airport which is in whole or in part under the jurisdiction of the Government Control Authorities. Where this jurisdiction of the Government Control Authorities does not apply, it is that part of the airport terminal building(s) with immediate access to the apron. In both cases, the air-side area is prohibited to the non-traveling public." (International Civil Aviation Organization [ICAO], 1953, p. 121)

For the specialized scholar, this is a fascinating attempt especially within the context of the early 1950s. The definition is rather ambiguous because instead of providing a single, standard case or condition, it provides two possible scenarios and one inflexible rule. The first part of the definition presents the airside as an area of the airport (without defining one) that is under full or partial control of the state. Meaning *any*, but the terminal. Surprisingly, the second part of the definition makes reference to the terminal and its zoning, but not the airport as a whole. So as a subtraction, it fails to define multiple areas that are implicitly left out.

Thus, if this speaks of areas and jurisdictions, it also implies borderlines and frontiers. Then, according to this rule, everyone traveling is allowed on the airside and therefore the apron, and it segregates only spectators and passenger companions.

There is also ambiguity in the understanding of what constitutes immediate access to the apron. Does it mean that there are no control points, or just doors and fences, or anything at all? One of the keys to understanding this definition is the fact that at the time every single passenger had to walk on the tarmac to enplane.

> Landside: "That area of the airport and buildings which is available to the non- traveling public." (ICAO, 1953, p. 128)

This definition is comparatively much shorter, meaning that whoever is in control of the landside—as here defined—is relatively irrelevant (implicitly, irrelevant for the state). Rather, the clear message from ICAO was that "everyone not traveling" was unwelcome within the state-controlled zones. The spirit of both definitions—understood as complementary concepts—is maybe clearer now; there is an underlying need to segregate, at this point at least, two different kinds of users.

From the above-mentioned definitions, one may infer that in this case, ICAO tried to use the landside–airside boundary as a super control mechanism. The creation of global institutions represents what James Scott (1998, p. 4) calls "high modernism," or a strong exercise of control through regulation and reordering through the coercive powers of the state (or group of states in this case). The physical reference in both definitions emphasizes the need to define material boundaries, though it leaves out room, for any non-physical dimensions of landside and airside. The text of *Annex 14* (ICAO, 1953) also includes the definition of an airside assembly area (airside waiting lounge; airside waiting room; embarkation room) or the "accommodation allocated to passengers on the air-side of the airport terminal building for their use prior to embarkation" (p. 122), and the concourse (common concourse; hall; landside assembly area; lobby; main concourse), which is "that area on the land-side of the airport terminal building provided for the convenience of passengers and the non-traveling public" (p. 122).

These two definitions speak less straightforwardly of the use of landside and airside, not as spatial labels, but as tools for segregation. It speaks indeed of areas which are separate, but as a norm does not let us know how or where to draw the line; what should designers do? The laxness of the definitions leaves clear that each airport or terminal layout would be resolved through *a per-case* negotiation. The standards are furnished with some drawings addressing specifically to the landside–airside as territories divided by an intangible boundary. The following illustration is perhaps one of the most relevant pieces of information regarding the topic of this study (see www.landsideairside.com).

This sketchy hand-drawn illustration shows us an example of how to treat the boundary as a material entity. There is a serious contradiction between this formula and my previous analysis of ICAO's definitions, as both the landside and airside are now invading the neutral territories of the terminal building. The use of a broken line suggests through points or lanes and avoids, as a representation, the implication of a partition wall.

In sum, the drawing resembles a hesitant approximation of *something* that is better to leave partially unclear or open to interpretation. (It is useful to remind the reader that these standards are international and are applied in all sorts of cultural contexts.) In addition, *Annex 14* mentions the "flows of load," referring to passenger and visitor patterns of movement. It demands that the "design of the terminal building on the landside should facilitate the disembarkation of passengers, under cover, and the unloading of their baggage from vehicles at the same point" (ICAO, 1953, p. 9). From this quote, it is clear that for ICAO the landside is the side of the building facing the road. In this same sense, the manual suggests that, "whilst it is not practicable to recommend a maximum walking distance within the terminal building, it is desirable that the distance which passengers have to walk from the land-side entrance of the terminal building to the air-side exit be as short as possible."

Moreover, "it is equally desirable that the route which the passenger has to follow within the terminal building be as straight and direct as possible" (ICAO, 1953, p. 18). Again, greater emphasis is placed on the existence of two faces of one building, one representing the entrance and one the exit. This in itself contradicts the glossary definitions discussed above (let us remember that before the 1970s, visitors' access was granted up to the doors that connected to the [often] lower apron).

As depicted in a 1953 photograph of the airport in Charleston, West Virginia (see www.landsideairside.com) the question of openness and free flow between people and aircraft remained paramount for airport designers. The waiting lounge portrays a relaxed landside–airside boundary, according to which the different layers of rooms, glass walls, promenade, terrace, apron, and airplanes are part of the same uninterrupted sequence. With regard to baggage handling, *Annex 14* sets the standard for locating domestic and pre-cleared baggage. The code recommends that the baggage claim be "appropriate to the arriving passenger flow process and with ready access on the land-side of the building to public and private transportation and that the unloading of departure baggage from vehicles should be off-loaded from vehicles under cover on the land-side at the same point as the passengers are disembarked" (ICAO, 1953, p. 29). At this point, ICAO literally portrays the landside as the road and the drop-off sidewalk, and this fact explains why so many planners keep using the same concept in the present day.

One more critical aspect addressed in the manual is that the word *segregation*—previously used by Froesch and Prokosch (1946) and in the multi-cited *Progressive Architecture* special issue on airport design of 1953—has been carefully replaced by the less-charged term, *separation*. The manual instructs designers on how to separate passengers from non-traveling people (such as relatives, friends, companions) as follows: In the case of international passengers, separation should be effected for departing passengers at the point where final personal control commences; this, where [outbound]

> customs clearance is required, is normally at the position referred to above as Customs Personal and Hand Baggage Control; where outbound Customs is not required it may be at the Currency or Passport Control point). It is important that adequate provision be made to ensure that the handling of passengers at the airport-handling counter (check-in point) is not impeded in any way, i.e. the airline handling counter should be isolated in a suitable manner. If this isolation is achieved by means of guard-rails, sufficient space should be allocated in front of the handling counter. In addition, arrangements should be made so that arriving passengers can join the non-traveling public at the point where they clear the final inbound control. (ICAO, 1953, p. 11)

This article clearly shows the difficulties of understanding and defining the materialization of the landside–airside boundary. Again, rather than being a norm, it is a set of design restrictions that are flexible and interpretive. But it is striking to see how ICAO needed to represent a physical frontier for each of the three options: at the outbound customs stand, at the currency or passport control point, and at the check-in counters using guardrails. Even for the early 1950s, norms that implied containment were hard to understand, and when put into practice, they usually eliminated security leaks. While *Annex 14* speaks of two possible configurations for an airport, neither is suggested as ideal and much less as a norm. The "finger system" is described as a layout, whereby protrusions from the terminal building extend into the apron area, giving access to aircraft parked on the apron, normally immediately adjacent to the finger. A finger can take the form of either an extension of the terminal building, a pier or a tunnel, or a fenced passenger walkway on the apron. When a finger takes the form of a pier, it can have one or more levels (floors). In contrast, the "open apron system" is defined as "a system whereby there are no protrusions extending into the apron area. Access to parked aircraft may be gained by walking or by motor transport and involves the passage of passengers, baggage, freight and other load to the parked aircraft across an area in which aircraft are being maneuvered" (ICAO, 1953, p. 62). In an "Open Apron System," walkways are defined by painted lines on the surface of the apron and aircraft are parked adjacent to the terminal building in more than one row, thus requiring that passengers and loads be moved through the first row to get to those beyond.

This austere classification is accompanied by a long analysis of advantages and disadvantages in where the finger system's outcomes are superior by far, with the exception that it does imply a higher initial investment. In short, this article shows the struggle between the old "Open Apron System" and the more modern finger system, and what takes place here is the fight between two epochs and two ways of understanding the landside–airside boundary. A good analogy for explaining the aforementioned is to imagine the piers as elastic extensions onto the apron, the result of which is now a much more reduced space and surface in proximity to aircraft but one that is, evidently, now much closer. In simple terms, it is bringing the terminal into the realm of aircraft forming a 90-degree or less angle with the concourse.

In conclusion, I suggest that ICAO experts' use of language throughout the manual evidences their own concern with classifying and reordering the parts that form an airport in the 1950s, and furthermore seeks to homogenize on a larger scale the routines and zones in order to standardize as much as possible. It is the need to break with the established jargon that perhaps explains the profuse use of some outdated and unconventional terms. In this context, the inclusion of landside and airside and, more specifically, the boundary worked well as a socio-technical tool that conveyed a much deeper thinking of the nature of the airport. Also, the acknowledgment of the landside–airside boundary as a tangible entity opened perhaps prematurely—though only partially—a discussion on how the boundaries are finally negotiated.

4 How Did Airports Gain "Technological Momentum"?

In August of 1940, *Architectural Forum* published a so-called *revolutionary* scheme for a new airport in Washington (Projected Airport by Fellheimer & Wagner Architects, L.L. Odell Aviation Consultant, 1940, pp. 85–87). Behind the scenes, L.L. Odell, a pioneer of airport planning, proposed a double-spike layout that surprisingly resembles the typical models of the 1970s and 1980s. Strikingly, the architects, Kellheimer & Wagner, who had extensive experience designing rail stations, had never designed an airport before. Odell, who also acted as a consultant for LaGuardia, quickly learned the failures of this airport and proposed waiting rooms that were shaped as long piers that sprang out of a central, circular concourse. This example is exceptionally relevant to understanding the early mentality of the planner. Odell's concept reveals many of his interests: He looks to reduce passenger discomfort by bringing the waiting areas closer to the planes, and he even proposes telescoping air—bridges in order to have an undercover approach to aircraft. However, the proposal was abandoned, and a scheme similar to LaGuardia's was built instead.[7] There was not enough credibility yet for the young specialist (Odell).

As I have mentioned before, just a few years after beginning operations, LaGuardia Airport became practically obsolete (or saturated and unable to cope with projected growth). This situation alarmed and alerted several politicians, who thought it impossible to keep investing such astronomical resources in infrastructures that could not keep

[7]The chosen scheme for Washington National was Howard Lovewell Cheney's conservative, post-Art Deco building. Nonetheless, the building was praised by *Architectural Record* as "perfect," "ahead of contemporary practice," and deserving a "five star rating in anybody's news" ("Washington National Airport," 1941, pp. 53–57).

pace with traffic growth and last even a few years in full service. Step by step, aviation slowly became a profitable business and a good alternative for long-distance travel, and traffic did keep growing. Thus, the need for a new airport in New York City became imminent; but this time, William Delano did not get a chance. As the Roosevelt period faded away, new political alliances emerged. Perhaps thanks to his connections to the higher spheres in Washington, the *socialité* and modernist architect Wallace K. Harrison was given the heroic task of planning a new airport for New York City (see www.landsideairside.com). Harrison & Abramovitz intended to reinvent not the airport but the way airport terminals might become smaller components of a larger system or collection of stations. Harrison baptized his concept airport city. The visionary planner imagined a huge, never-seen-before elliptical central island, with several air stations or terminals arrayed around its perimeter, each devoted specifically to house just one airline (also called unit terminals). This concept survives in just one paper sketch that captures one of the most important airport-planning innovations of all time.[8] This fresh planning strategy proved to be much more long-lasting than any prewar examples such as LaGuardia. However, each building in particular was again thought of as a terminal and contained somewhat limited frontage length toward the apron. The deep platforms and new airport layout particularly favored the possibility of rotating the terminal on a ninety-degree angle in order to work as a pier. The New York-based firm Skidmore, Owings and Merrill (SOM) introduced this idea at the International Terminal (one of the nine planned by Harrison), which was placed perpendicular to the street. SOM gained an increased aircraft parking capacity but did not offer anything new in rethinking the airside–landside boundary. This is a clear example of how airports formed a sort of linear progression: Designers would pick the virtues of the previous scheme, work out some smaller improvements, and then devise a slightly different layout. Unfortunately, the proliferation of similar layouts left unbuilt a number of visionary schemes, such as the 1954 design of Paul Gerhardt for the Chicago Municipal Airport. Often these missed opportunities reflected the rule of the game, which was not

[8]See the first definition of "unit terminal" ("Planning of jet airports," 1960, p. 169).

efficiency but political value. The existence of a sort of mainstream tendency in airport planning during the early postwar period was probably co-responsible for the serious bottleneck congestion in Western airports around the mid-1950s. In the following pages, I do not critique this technological pattern but instead introduce the socio-technical facts that established the context of an atypical and unexpected rethinking of the airport as a whole.

In the book *Technology and Urban Transportation* (1965), authors Meyer (Harvard), Kain (Rand), and Wohl (MIT) affirm that "Americans are becoming an urban people and at least one consequence of this process of urbanization is that transportation of people and goods into and out of cities increasingly has become a matter of great public concern" (p. 1). Although this claim is not necessarily true, it is interesting to see how around 1960, regional transportation reached out to a larger public, both urban and suburban. In these upcoming years, airlines began to surpass the airport's capacity to service the growing flight demand. According to *Architectural Forum*, most of the issues behind the aviation boom "stem from the fluidity of the airline business itself. The civilian airport-building program, held up during Second World War, faced a greatly expanded interest in air travel in the postwar period. Traffic- growth estimates have nearly always been conservative. Result: the program has been—and still is—trying to catch up with demand" ("Airport scramble," 1956, p. 131). The so-called Jet-Age airports ("Jet-age airport," 1959) rapidly turned out to be more of an "airport scramble" ("Airport scramble," 1956, p. 117). Available facilities were insufficient and produced long lines at check-in, and often featured low ceilings and insufficient air circulation. But the real nightmare came when passengers were pushed to walk extremely long distances to board or set down.

Again, *Architectural Forum* described the true situation for users: "in the air the customer may be treated like a king. But on the ground he still has to fight his way through a chamber of horrors. Sheer size and the coming of 160-passenger jets call for new thinking about air terminals" ("Airport scramble," 1956, p. 117). When *Architectural Forum* published the article "New Thinking on Airport Terminals," the editors subtitled a section "Which usually turn out to be civic monuments when they should be pipelines for passengers" ("New thinking of airport terminals, 1956), in clear reference to spectacular terminals such as TWA Terminal in New York's Idlewild or Saint Louis's Lambert Terminal and others. This critique alone captures the frustrations and goals of airport planners; airport developers wanted fewer architects and more technical experts (planners wanted the airport to work as a machine). In contrast, in a 1956 pamphlet London's airport describes itself as "Passenger's Progress" (Chandos, 1956). The planning solution at the self-named "New World Air Centre" introduced twelve "channels" of flow, or processing lanes, which guaranteed agility and rapid transit through the terminal and resembled, at least in concept, a pipeline. In a compelling photograph, we can see how baggage was also "channelized," thanks to a spectacular (at the time) mechanized conveyor, which was proudly displayed and separated from viewers with a fence (see landsideairside.com). The channels protruded from the parallel airside façade, ramping all the way down to the apron, where pilots and drivers parked planes and buses as close as possible to avoid passenger exposure to the elements ("Airport scramble," 1956). The study makes a tough critique of American airports in general and calls its glazed exterior channel an airside gallery.

As the number of airplanes grew in the Western hemisphere, so did the number of passengers, in direct proportion; as a result of the definition of landside–airside boundaries, new airports faced new problems that required new solutions. In the USA, according to the President's Airport Commission, new facilities were "local problems" that needed to be designed while keeping in mind enough airport versatility to cope with technological change and social and political requirements and to maintain a harmonious relationship with their surrounding environment (President's Airport Commission, 1958, p. 41). However, after the introduction of new aircraft technologies such as the jet engine, technology brought new problems such as noise and fumes. This adverse condition formed a sort of noisescape.

As a nationwide response, "the FAA administrator was instructed to form a group composed of noise experts, lawyers, and traffic controllers to address the immediate issue" (FAA, Executive Director, 1962). After receiving endless passenger complaints, ICAO decided to recommend as "a general principle that the Administration should study the serious effects of noise, fumes and dust with regard to personnel, passengers and visitors at an airport. This problem is particularly acute on the apron side of the buildings. It is most important that the buildings be provided with adequate sound-proofing and air conditioning (including temperature, humidity, purification, and volume change control). Every practicable means of abating or eliminating noise, fumes, and just should be examined with the view to possible incorporation in the design, layout and construction of the building(s) and apron(s)" (ICAO, 1953, p. 5). This rising public pressure had a tremendous impact on the conception of the landside–airside boundary after the mid-1950s. Now buildings were required to be soundproofed and enclosed from the external noise conditions. This situation set out an environment similar to Thomas Hughes's example of the Americans' early electrification systems, for creating technological momentum. (Just as, in order to solve the problems of the postwar era, knowledge institutions had to be created and new incentives for inventors and engineers made available to find workable solutions.) Users had finally challenged airport engineers, planners, and designers: How could they enclose the passenger's path to the cabin of the aircraft? The race for invention had begun, and most aviation companies had to put together design and engineering teams that were looking for practical solutions to this problem. Among the noteworthy innovations of the time, the early use of air bridges between terminal buildings and planes was a radical departure from the romantic borderline.

In 1951, an early version of the "Loadair System" developed by Whiting Co. was planned to connect planes to the pier in a very literal way, but it complicated things by trying to draw planes on dollies running on sunken tracks. The following year, a simplified variation of the "Loadair System" was placed in experimental operation by Avianca Airlines in Colombia, South America (see www.landsideairside.com) ("Loading made easier," 1952, p. 26). Prokosch (1951) describes in *Architectural Record* what he calls mechanical docking: "the device was similar to the locomotive transfer table in general use in railroad repair shops. It consists of a pair of tracks each of which supports a dolly. The airplane taxis onto a dolly with each of its main tires, and is then brought toward the loading gate by means of electric-powered cables. Projecting from the terminal is a fixed two-story finger with, at its outer end, a short ramp which can be adjusted to varying cabin sill heights" (p. 115). Its original purpose was to expedite cargo, although soon after, the company saw the convenience of using the system for passengers and could avoid their exposure to unfavorable weather. Following the same principle, companies like Lockheed Air Terminal Inc., started developing prototypes, although it was not until 1956 when Universal Fabricating Co. developed the first commercial air dock and tested it in November of the same year in Travis Air Force Base in the USA. Just one year later, in 1957, the race began when United Airways developed the "Air dock" prototype (which was a refined variation of the above, but intended for passengers). In 1958, according to the Chicago Tribune, Lockheed Terminal Inc., invented the "Aero-gangplank" a device similar to a telescope. The artifact was tested and installed in one of the piers of Chicago O'Hare Airport in March of the same year (see Fig. 2).

It is also interesting to mention that in the following year, during the launch the first DC-8 flight of Delta Airlines in June, Delta used one of the oldest jet bridges known, although it only spanned no more than nine feet from the tip of the pier and gate 11. Just a month later in the same 1959, Pacific Iron and Steel, Co. Ltd. installed a new jetway prototype in San Francisco International Airport. By the end of the same year, JBT Co. installed the first patented bridge called "Jetway 1" at the same airport with great success. The jet bridge marks an important shift in the history of the landside–airside boundary (see Fig. 2 and find striking photographs of all the aforementioned in www.landsideairside.com). With the exception of smaller regional airports, the transit from the building to the cabin was finally isolated, but not without complications. Designers had to develop a device capable of adapting to the rapidly changing heights of airplanes. The jet-bridge is a typical example of a "heterogeneous technical object" that has been constructed by diverse forces (Akrich, 1992, p. 205). It is my contention that the invention of this artifact embodies two kinds of "actorships": First, it responds to the interests of passengers (whose complaints were impossible to ignore), and second, it depends on the requirements of the machine

Fig. 2 Readers should consider the technical difficulties that different aircraft heights pose to engineers. The so-called gangplanks needed to mechanically adjust the levels and at the same time keeping the slopes comfortable for passengers. Notice the invention of the "telescoped" system. Fenn, Al (1958). *A scene from O'Hare Field, at the United Airlines landing strip, as passengers walk down the interior of the gangplank which will mount between plane & building* [digital] (Reproduced with permission The Life Picture Collection/Via Getty Images)

(the diverse shape and sizes of aircraft complicated for years a possible solution). I suggest, therefore, that a more current reading of these artifacts should include these formative features. Although the airport configuration during the 1950s became a clear example of technological momentum, few designers responded to the growing complexities of the landside–airside boundary. Most of the designs opted either for lateral growth, producing incredibly long buildings parallel to the runways, or for the newer perpendicular piers recommended by ICAO. These adaptations and new configurations often left aside the fact that social aspects now would play a role in direct proportion to both traffic growth and local culture particularities. But experts as well as designers were tempted to predict what was coming in the next few years, and they didn't miss the mark by much. For example, Prokosch (1951) foresaw the "mechanization of docking" (p. 115), which he considered not economically viable.

The author and airport planner also thought that "for many years the bane of passenger handling—at least from the passengers' viewpoint—has been the fact that loading an airplane cannot be accomplished with protection from the elements. Extensible canvas awnings were used at a few airports with meager success. The ubiquitous umbrella almost lent a touch of comedy to the loading process as passenger and agent leaped among the puddles in a dash toward the cabin door" (p. 115). On another front, American Airlines engineer E. W. Fuller (building on the European bus model), proposed to transfer passengers on double-decker buses, which could load at ground level and feed passengers at the same height of an aircraft (see www.landsideairside.com). His proposal was presented at the Society for Automotive Engineers in 1952, but was later dismissed ("Airport bus," 1953, p. 69). Fuller's design was called the "mobile gatehouse" and curiously somehow anticipates Saarinen's solution for Dulles.

Nevertheless, technological change was far from homogeneous. Two terminals that went far from the *planning mainstream* became symbols of their time: TWA Terminal in New York's Idlewild, and Lambert Terminal in St. Louis. In 1956, just a few months after Minoru Yamasaki's structurally monumental terminal opened to the public in Missouri, Saarinen was commissioned to design Terminal 1 for New York's Idlewild. His approach, in his own words, was to accomplish what Yamasaki left as "something to be desired as a flight terminal" (Temko, 1962, p. 46). Amid the rapid changes of the time, the iconic TWA terminal in New York became an inspiration model, where metaphor and symbolism played together with a futuristic aesthetic of connecting tubes and bridges. Leaving the springy lawns of the airfield aside, TWA offered stunning views through its monumental elliptical rose window. No doubt, the legendary terminal became for many the technological cathedral of the future. Pulitzer Prize winner Allen Temko (1962) described right after Saarinen's death why TWA became so important: "Individual man, in the symbolic personage of the traveler, was given consideration, on the whole, than were machines and corporations" (p. 46). However, few know that the main forces behind the grandiose design were the airlines and not the city's administrations (as is commonly thought). When Wallace K. Harrison individualized terminals in his plan for an airport city, he made them exclusive to airlines, thus triggering corporate competition and the use of architecture as a means of large-scale branding.

5 Toward the Reinvention of the Airport

By the time FAA was established in 1958, airspace had been successfully regulated and normalized in the USA; thus, aviation marched steadily toward its consolidation as a large technological system. But during the roughly twenty years that separate the planning of LaGuardia and the airport congestion crisis around 1960, the traceable changes at the landside–airside boundary were relatively gradual. In fact, it was the technological continuity and its own consolidation into more formal knowledge (the planning profession) that kept the debate on the landside–airside boundary at its lowest. However, social pressures continued to rise as congestion started to hit most large-scale airports across the USA and Europe. The international scene of airport planning was suffering from similar problems across the board, and still under the not-so-distant shadows of the Second World War, the nationalistic temptation to form new systems from zero was certainly strong.

In the USA, after the birth of the airport city at New York's Idlewild, President Eisenhower's special Airport Commission of 1958 reported that the future may "require super-airports and at least six of them… they should be located at a considerable distance from metropolitan centers and high costs could be anticipated" (quoted in Chandos, 1956, p. 27). This piece of advice certainly paved the way for a possible reinvention of the airport as a system.

In the fading days of the CAA, presidential advisor General Elwood R. Quesada took over the FAA and the tortuous mission of furnishing the capital city with a brand-new airport. Quesada put himself at the forefront of a team of FAA experts, plus Ellery Husted, a reputed local master planner and personally took over the task of finding the right land plot for the new airport (Geraci, 1965, pp. 15–16). In 1958, once the preliminaries were ready, New York's firm Ammann & Whitney, a high-caliber engineering company that held in its resume nothing less than the George Washington Bridge, was awarded the contract. Quesada "zealously insisted that the passenger should not receive short shrift; instead the objective was to get them through the airport as safely and expeditiously as possible" (Geraci, 1965, pp. 15–16).

During the previous decade, the voice of the passenger was mainly heard through the press and letters to the authorities. In 1961, travel journalist Robert C. Ruark ironically described how "one of the more desirable aspects of the modern airport is its capacity for serving as a training camp for athletes. I don't take the long healthful strolls in the country anymore, I just travel by air" (Ruark, 1961). For passengers, the experience of traveling became gradually less pleasurable. In the sequence of a departure flow, for example, passengers would have to make long lines at short drop-off curbsides, be rushed by traffic officials, and again make new long lines at the check-in counters. Once documented, users and their companions had to walk down long aisles to reach their departing flights. In the end, departure halls were totally overcrowded because passengers and companions were not yet separated. The last nuisance was the imperative need to use stairwells (USA) or ramps (Europe) and to walk on the tarmac regardless of the weather conditions.

By the end of the 1950s, one more inconvenience was the exhaustingly long boarding time. "It sometimes seems to take longer to get to an airplane than it does to fly 300 or even 50 miles," passengers complained to the *New York Post* in 1959 ("Jet-age progress," 1959). Distance and delay certainly created a paradox. A year before, others papers such as the *Sunday Star* in Washington chronicled how "for the commercial plane passenger, the average hike from ticket counter to plane was about 1/5 of a mile. Passengers transferring from Continental to American Airlines at the new Chicago's O'Hare jet airfield must walk 9/10 of a mile from arrival to departure points" ("You ride to plane in mobile lounge," 1962, pp. 36–37).

But it was the FAA who finally decided to remediate the crisis. On the one hand, the agency had to cope with increasing demands from traffic and thus the airlines and the travelers. Through its Public Affairs newsletter, the FAA (1961b) declared that "the magnitude of Dulles International Airport and the tremendous air traffic to be handled there in the foreseeable future caused the FAA to search for new methods of solving the passenger handling problems facing most airport terminals today. The Dulles terminal will provide the necessary facilities required to streamline passenger handling for the jet age" (p. 2). At the time, airport facilities were often patched with temporary solutions that only remedied the issues for short periods. Thus, the FAA was particularly "concerned with the tendency for airports to grow by the construction or extension of finger systems. Finger systems cause tremendous inconvenience to passengers when enplaning, deplaning or attempting to transfer from one airline to another."

But not all passengers agreed, especially because now they were less exposed to the elements. For the FAA (1961b), however, "the conventional finger system makes the passenger walk long distances through the fingers and additional distances to the plane, exposed to wind, rain aircraft noise and propeller blast from adjoining gates. This method often requires long and intricate aircraft taxiing" (p. 3).

In addition, the agency did not favor the loading bridge system used at some airports because it also "involves long walking distances thought finger buildings, although all on one level. It does, however, keep the passenger sheltered at all times." This is a very important assumption that the FAA makes with respect to the newest technology on the market. It discredits the system based on the above reason and the argument that they "also require costly, intricate aircraft taxiing and demands precision positioning at the gate" (FAA, 1961b, p. 3). Nonetheless, agency officials thought that "by contrast, the normal highway type bus system in use at London, Frankfurt or Amsterdam cuts down the passenger walking distance" (Chandos, 1956, p. 27) (also known as channels in the UK). This motorized solution allowed the British to operate a very dynamic ramp at Heathrow where many aircraft could be parked remotely and perhaps have smaller satellite facilities as support. However, officers complained that the bus would "still require changes in levels and the passengers are exposed to weather, noise, blast and fumes during the transfer from bus to plane" (FAA, 1961b, p. 9). So in the end, the FAA was in need of a new passenger-handling concept, not just at Washington Dulles, but nationwide.

In the case of Dulles Airport, what had to be finally reformulated was not only the object (terminal) or the system (airport) but its meaning (e.g., it would be difficult to reinvent a seaport from scratch without understanding the cruise ship and the ship passenger, plus their in-between relationships). In this airport, those relationships of mediation would be part of a negotiation process in which different actors pursue different interests.

The outcome would be a new kind of landside–airside boundary configuration, one never seen before, resulting in a whole new kind of airport. This section traces those voices (FAA, designers, planners, engineers, airlines, passengers, pilots, journalists, administrators, government officials, and others) and analyzes their role in negotiating and eventually materializing a reinvention of the airport.

6 The Birth of the Mobile Lounge

Perhaps the history of Dulles Airport should be divided between the planning and design phases. Apparently, this would be consistent with the fact that after the consolidation of planning as a profession between the 1950s and 1960s, both activities would split up a good deal of the work. However, there were unique conditions behind the initial organization of Dulles; the client concentrated most of the load of both specialties in the hands of the architect. Charles Landrum, a pioneer airport consultant with Landrum & Brown, acted as the external consultant, but the FAA left him little room to maneuver. Remarkably, this was not necessarily the case in practice, since Saarinen split his staff into two main branches: field investigators and designers. Therefore, an integrated account would be historically imprecise.

Although it is General Quesada himself who wanted to hire Saarinen (Geraci, 1965, p. 16), in the case of Dulles, the pressures to deliver a scientifically sound study (that could justify a new airport concept) initially relied entirely on Ammann & Whitney and not on the architect. I should point out again that FAA contract #33777 mentions neither

Saarinen nor any of the other specialists. Thus, the responsible party must have inflicted a tremendous pressure on the subcontractor. By the end of 1958, Saarinen put two teams to work in parallel. One key condition was necessary in order to come up with a whole new idea for an airport. According to Hughes, "the system builder needs to direct a process involving problem identification and solution as an idea" (Hughes, 1983, p. 19). As I showed in the previous section, the FAA provided a very solid problem identification (from users mostly, rather than from any theory) and eventually pointed toward the most desirable solution. For the FAA, the European bus system was perfectible; thus, the message was sent straightforwardly, and Saarinen was expected to come up with an integral solution.

During the first year of preliminary studies, Saarinen designated two teams, which were put in charge of carrying out two central tasks: a comparative analysis of airports and a study of the most common airport problems. To come up with a comprehensive comparative table, the Finnish architect scheduled a number of trips throughout the world in search of the most modern airports and to research passenger complaints and general airport issues, Kevin Roche—one of his closest collaborators and his eventual successor—was put in charge of surveying airports worldwide in order to list all sort of problems, layout weaknesses, delays, and passenger complaints. Basically, the surveys' purpose was to feedback architectural schemes toward an airport that would fulfill Quesada's main concern: to expedite transit through the airport at any cost.

Thanks to a generous budget, employees were sent in all directions to "time" and "evaluate" airports. In 1962, *US News & World Report* published a special reportage on the planning of Dulles. In the article, the editors describe how Saarinen organized the task: "before he drew a line, he sent out teams with counters and stop watches to see what people really did at airports, how far they walk their interchange problems" ("A look at the newest thing in airports," 1962, p. 78), as he put it. He found that as airports grew and jets consumed more and more space (a 200-foot parking diameter), travelers had to walk further and further along miles-long 'fingers' to get to them" (p. 80).

The Washington *Sunday Star* reported that "Architect Saarinen, trying to find a better method of airport transportation, bought stop watches and counters and sent men to several airports to clock passengers time between flights" ("You Ride to Plane in Mobile Lounge," p. 37). In the firm's logbook, several airports are listed for testing. Visits were organized to Zurich, Philadelphia, London, Sydney, Montreal, Newark, Dublin, Anchorage, Charleston, Edinburgh, Seattle, Rio de Janeiro, Chicago, Long Beach, and Burbank, among others. The envoys reported back to a table of solely quantitative data, which included relationship tables of distances, times, basic characteristics (finger, pier, satellite systems, etc.). Those tables were handwritten on paper, sorted, selected, and later photographed.

In the meantime, Mr. Roche made slow progress in his personal striped notebook, where several pages show small sketches of traffic flows with passenger patterns in the shape of arrows pointing in different directions. His notebook also evidences the lack of an appropriate requirement program; Roche himself was gathering a long list of locales and facilities that could or should be included. Roche and his collaborators were in fact dealing with the geometries of aircraft taxi and parking movements at the ramp. In his notes appears the name Froesch & Prokosch, and many of the requirements and suggestions are direct transcripts of the original airport-planning treatise. This may explain the total absence of the terms landside and airside from all documents and technical drawings.[9]

Even the word airside was replaced on many documents with the unusual aircraft space. Among the various diagrams and drawings produced by the architect, several propose separation of flows in sections; others, slightly more elaborate, organize the different elements of the program, similar to TWA in New York. These sketches illustrate architectural parties of internal and external organization around a major

[9]The original Masterplan submitted by Ammann & Whitney, and Saarinen, mentions neither the landside nor the airside (Ammann & Whitney, Eero Saarinen, Burns & McDonnel, Ellery Husted, & Burhan Kelly, 1960).

central hall called "the waiting room" that is similar to Saarinen's approach for TWA in New York; from this room piers sprung to a parallel loading pier, also emulating LaGuardia's Skywalk.

The process I have discussed so far speaks of little progress in terms of laying out a new airport, but enough data were collected to inform two key charts: A table called "Passenger Walking Distances" (see www.landsideairside.com) illustrates how large airports such as Chicago and Los Angeles had very long walks, 6200 feet and 6500 feet, respectively (under their worst possible condition), but in general the means are close to 2000 feet in walkable length. The second chart shows a curve in which the "average" finger scheme for Dulles would require that passengers travel forty-five minutes from check-in to departure ("Time study—Mobile lounge," 1959). In the final scheme, hypothetical times and distances were added to these tables and compared as technical evidence. In parallel, Saarinen was in charge of the design aspects. His immediate reaction was to use the very standard method of analogues. This model, popularized through architecture schools since the days of the Beaux Arts, calls for a compilation of similar buildings that resemble the genre, the program, or the purpose of the target building. Once the analogues are gathered, there are critique sessions in which the lead designer speaks of the pros and cons of each scheme (or experience) and tries to unveil the "metis" (Polanyi). The trained eye of the master-lead designer ponders the previous layouts in terms of style, efficiency, structure, aesthetics, abstraction, economy, responsiveness, buildability, context, character, and originality.[10] Above all, a design is a hypothesis that the creator tries to elevate to the category of a paradigm. Extrapolating from Kuhn's model (Bird, 2011), the hypothesis must seem better than its competitors and gain acceptance from expert groups in order to be considered paradigmatic.

Saarinen traveled extensively, as his letters and personal documents show, in search of the best and most modern airports. On some of his visits, he announced himself as a sort of ambassador and requested

[10]Paraphrasing Lackney (1999).

technical visits.[11] In other trips, such as to London, Sao Paulo, Athens, Rome, or Zurich, he thoroughly toured the facilities. From his private collection of hundreds of slides, it is evident that his interest was eminently architectural and not technical in terms of airport planning. It is particularly revealing that on two slides special marks cross out the titles; these photographs bear a striking resemblance to certain back angles of Monco & Luccichenti's Fiumicino Airport and the final version of Dulles.

Back in America, he traveled to the west to visit the spectacular Lambert Terminal at St. Louis International, designed by his direct competitor, Minoru Yamasaki. Of all his trips, this one perhaps became the most influential during the design process. Back in Hamden, Saarinen started a series of sketches in which he became seriously concerned with the structural shape of the possible terminal building. The drawings depict different geometries and possible organic shapes for spanning a nave. On a more finished sketch, the architect shows how the main façade of the building forms a monumental colonnade and a long car drop-off sidewalk.[12] Here Saarinen shows his interest in what historian of engineering David Billington (1983) has called structural art, which is understandable if one compares the very high standards set by the beautiful concrete vaults of the St. Louis airport.

Speaking of system builders, Hughes (1983) suggests that "the first expression of an idea often appears as a drawing in a notebook or a scrap of paper" because inventors are frequently visual (p. 19). In Saarinen's sketchbook though, there are no accounts of the landside–airside boundary concept of Dulles. Rather there are multiple roof alternatives, vaulting systems (some similar to Nervi's), colonnades (some derivative of the same Yamasaki), and structural shapes. His concept

[11]Saarinen often asked Charles Landrum, his famous airport planner and other governmental officers to write in his behalf in order to receive special credentials in the airports he was touring. For example, in a letter directed to Pierre Cot, Paris Airport executive director, he expresses his interests in "airport problems" and asks for assistance in going through the "magnificent new facility." Saarinen, E. [Letter to Mr. Pierre Cot. May 18, 1961. Saarinen, E. (1961, May) 593. Box 8, Doc 95]. Collection Manuscripts & Archives, Yale University.

[12]Saarinen, E. 3080–3133. Box 463, doc 59. Collection Manuscripts & Archives, Yale University.

does not yet reveal the reorganizing vision of the airport as a whole. But following my extrapolation of Kuhn's model, many pieces of the puzzle now fit. Yamasaki's scheme had two opportunities for improvement:

- Planning: Replace the finger scheme.
- Design: Integrate the split vaults in one single nave (in the ideals of modern architecture the sublimation comes from achieving full homogeneity or a "universal space," completely lacking of historical or cultural references).

It is through this possible model of competition, I argue, that Saarinen resolves the riddle: In terms of planning, he follows the bus model of London Airport (with which the FAA sympathizes), but in order to fix the variations between the aircraft and the building (the nature of the variable machine versus the stillness of the terminal), he turns it into a technological challenge. Innovation passes now to the hands of Chrysler, the quintessential American company of the 1960s. Saarinen has given them the idea, a simple one perhaps, but now the pressure is on the fabricant. This may explain why Chrysler Corporation cites both the design and construction of the Mobile Lounge as theirs in several advertisements of the time.[13] Freed from the planning straightjacket, Saarinen tackles the design aspect and uses the same structural system he used a few years later for the Yale University Ice Rink. The roof will be strung from cables, producing a support-less open plan. "It will never be hard for critics to find basic decisions which irrevocably shaped the form and movement of this jet-designed field. CAA selected the runway pattern it wanted for south-to-north and north-to-south operations with jets landing and taking off simultaneously. It adopted the Mobile Lounge concept to move passengers between far-flung ramp positions and a centralized, tight-knit terminal building (see Fig. 3). Virtually nothing that has been done since these two decisions were made—long before first contracts—has been unaffected by them" (Dunbar, 1960, p. 37). This fascinating dichotomy may be understood through the lens of Pinch and Bijker's former idea of

[13]Eero Saarinen Collection (1938–1962), *Printouts of advertisements by Chrysler, publicizing the virtues of the Mobile Lounge (1961–1963).* Manuscripts and Archives, Yale University Library.

the technological frame. More specifically, for Bijker (1997), technological frames are built up as an interaction around an artifact begins, and once constructed, they shape activity. In this case, Saarinen's suggestion of a mechanically adjustable bus (Mobile Lounge) depicts an engineer's response to the key problems the artifact is sought to solve; therefore, it is "imprinted" (Akrich, 1992) with very specific routines and codes. In contrast, the terminal is a neutral artifact, devoid of meaning or specific purpose. It could be a terminal, but also anything else that may need a shelter. Thus, Bijker's idea of the technological frame helps explain why the interactions between the relevant actors (Saarinen, FAA, Chrysler, Ammann & Whitney, Landrum, Quesada, Burns & McDonnell, Ellery Husted, plus passengers, the press, politicians, and so forth) spin around the Mobile Lounge and not the terminal itself. It also clarifies why this technological object follows a process of substitution for previous landside–airside boundaries and how, in doing so, creates a circle of activity that may be understood as well as a continuous negotiation.

However, things were far from decided. According to an article published in *New Yorker*, "to sell his scheme, Saarinen had his long-time friend Charles Eames make a film on the trials of an air passenger, in which one background noise was the squeak of shoes. Washington officials were sold last week" ("Jet-age airport," 1959, p. 18). In fact, the concept was immediately welcomed by many, even in its most preliminary phase. Journalist Brad Dunbar said in *Airlift* that "snuggling against gates built two-abreast in each bay on the field of the terminal will be the Mobile Lounges. Whether 'detachable terminal fingers' or glorified buses on stilts, they will be the true key to Dulles International Airport" (Dunbar, 1960, p. 37). However, much was at stake now, and the final efficiency of the new system was hard to predict. In a master plan preliminary report, Ammann & Whitney and Saarinen presume that some of the assumptions on the plans for the airport have been based will inevitably change, so they ask the FAA to develop a mechanism for adjustment (Ammann & Whitney et al., 1960).

But on other fronts, the first design progress reports triggered new discussions with new interlocutors. The airlines, for example, categorically rejected any responsibility for supplying a traffic quota or for paying any of the construction costs of Dulles. The airlines were against the

Fig. 3 A new kind of apron reinvented for the future. Korab, Balthazar (1963). *Aerial view showing mobile lounges* [digital] (Unrestricted use. Balthazar Korab Archive at the Library of Congress)

Mobile Lounge from the beginning, because the estimated costs were greater than those of the finger type operation. They made an offer of $360,000 per year of use. Negotiations to discuss the airlines' participation and project scheme lasted six weeks.[14] The Association of Pilots &

[14]Bacon, R. F. [Director of the Bureau of National Capital Airports]. (1962b, April). [Letter to the Administrator of the FAA Memorandum No. 7941]. From the U.S. State Government.

Aircraft Owners (AOPA) sent multiple letters to the FAA and published others in their bi-monthly newsletter claiming that:

> individual and corporate taxpayers in the 50 states are committed to pay some$150 million for the land, construction cost, landscaping and access facilities of Dulles Airport. These taxpayers and aircraft users are expecting a regional jet airport that will meet the requirements of those who desire long-haul international or domestic jet transportation. They expect airport designers to lay out a plan that will provide equal convenience at the terminal building for seat-holding airline patrons, no matters what modern conveyance they use in arriving to go aboard any airline transport. They want no discrimination with reference to loading accommodations, between the passenger who arrives by air taxi and one arriving by highway taxicab. They expect design experts to select adequate space near terminal building doors for both the rental airplane user and the car user. They will want a prospective airline passenger who arrive in a private or business airplane to step out of his vehicle as close to ticketing and baggage services as will the passenger who arrives in the family's Chevrolet or his company's Chrysler. (Thompson, 1962, p. 1)

But the FAA felt that the Mobile Lounge concept was a model that could be replicated in the future, perhaps even a typological building. And perhaps Chrysler saw big business in the future (see Fig. 4). For the FAA, "every possible method was explored to increase passenger comfort, convenience and to reduce the walking at Dulles International Airport since it is the first airport being designed for the jet age from the ground up" (FAA, 1961b, p. 3). In the same vein, in a letter referring to an administrative report to the US Congress, Acting Director of the Bureau of National Capital Airports Robert F. Bacon assured Congressmen M. Gerhardt that the finger system was not only outdated but, relying on the studies made for this project, economically less costly. Bacon affirmed that exhaustive studies under his supervision verified that the Mobile Lounge is far a superior alternative to the satellite and finger systems.[15] Others published: "So it is encouraging to

[15]Bacon, R. F. [Director of the Bureau of National Capital Airports]. (1962a, March) [Letter to U.S. Congressman M. Gerhardt].

Fig. 4 Commercial postcard showing how the passenger's upload to the aircraft now takes place in the middle of the apron. Korab, Balthazar (1958). *Mobile Lounges* [digital] (Unrestricted use. Balthazar Korab Archive at the Library of Congress)

learn that at least something is being done about shortening the time it takes to get aboard after you have reached the air terminal" ("Jet-age progress," 1959). Both the planning and the design were finished and a supportive environment was being built.

The FAA internally published four reports called "fact sheets," the purpose of which was to homogenize the information that had to be communicated to the exterior. The sum of these reports forms what

I call here the official version. For the FAA, "the terminal consists of a single compact building. It is therefore radically different from the usual terminal building, which sprawls out with fingers extending to distant plane positions. This compact terminal was made possible by the adoption of an entirely new method of passengers movement from terminal to parked aircraft—the FAA Mobile Lounge concept" (FAA, 1961a, p. 2). For the FAA, Dulles represented an "entirely new method of passenger movement from terminal to parked aircraft" (FAA, 1961a, p. 3). Thus, what spanned the gap between previous airport examples and Dulles were the proposed innovations at the landside, the airside, and the boundary that divides them. For the Association of Pilots & Aircraft Owners (AOPA), Dulles "was designed to eliminate the walking distance and time delays that have steadily increased at the conventional finger type airport building. When Mobile Lounge vehicles are used, the large jet planes do not come up to the terminal, but are serviced on a remote central service apron between widely spaced parallel taxiways and runways" (p. 3).

This terminal building area arrangement permits the passenger arriving by automobile to unload directly in front of the airline which serves him. Just inside the entrance doors are the ticketing and baggage check-in counters. The Mobile Lounges wait in docks on the opposite side of the building, as little as 150 feet from the doors. "Between the check-in counters and the docks, in an appropriate location, are consumer service or concession areas where newsstands, shops, restaurant, coffee shop, and other public facilities face the field" (Thompson, 1962, pp. 3–4). The FAA (1961c) maintained that "from an airport management standpoint, it allows maximum operational flexibility of the airport by placing aircraft operational facilities separate from passenger handling facilities and placing each in its own uncongested area" (p. 4). Again the agency emphasizes another of the major changes at the landside–airside boundary. In its official fact sheet, the FAA (1961c) highlights the new features of the concept in relation to aircraft and taxiing: "Another distinct advantage of the Mobile Lounge concept is that the passenger terminal can be concentrated in a single structure rather than in the sprawl of numerous buildings now found at most major airports. In addition to simplifying the passengers' problem of enplaning and deplaning, the new concept requires a minimum of aircraft taxiing to and from gate positions. This is extremely important to the big jet transports since they weigh upwards of 200,000 pounds and consume large quantities of fuel while taxiing" (p. 4). The official description of the terminal continues: "passengers entering the lounge at the terminal will pass on either side of the driver's compartment and enter the spacious lounge.

Music will be piped in from the terminal building. There will be carpeting on the floor, soft pleasant lighting overhead and tinted windows to filter light from outside" (p. 9). Once passengers approach the gates, they will board a Mobile Lounge. "One end is designed to connect to the terminal building utilizing two wide entrances. The other end is designed while all four-engine aircraft are at use today, and in the foreseeable future by domestic or international carriers. This connecting device will permit an easy transfer of passengers between the Lounge and the aircraft at the level of the aircraft to the building at which point it automatically becomes a comfortable waiting room or lounge of the terminal building" (p. 2). In the Mobile Lounge, "the passenger compartment will accommodate a total of 90 passenger with 73 seated" (p. 9). Access to the local services or feeder planes is by a central wing that projects out beyond the main concourse of the terminal. "The upper story of this wing will house a beautiful dining room commanding a full view of the jet gates and the landing field area. Also the roof of this wing will serve as an observation platform" (p. 6). In conclusion, three aspects stand out from the "official description" of the airport: First, the FAA discredits the "finger" option; second, they make clear that the airport is a "radical departure" from the past; and third, they emphasize a new level of passenger comfort throughout the complex.

7 Materializing a New Landside–Airside Boundary

The new landside–airside boundary in Dulles rephrases our historical dialogue with aircraft. Although a technologist may trace some continuity and branch evolutionary patterns between LaGuardia in 1939 and Dulles in 1962, it is not until now that the boundary allows a full revision and questioning, from top to bottom, of the airport's past. This in-depth revision shows how the increasing tensions within closed systems result in eventual crisis and thus a reformulation of the most basic terms becomes necessary. In the end, Dulles does not change the airport at all; the essence of the airport remains untouched. Planes land and depart, people arrive by car, bus, or on foot; somehow they enplane and deplane; sightseers awe, employees work, baggage is loaded and unloaded. Although multiple voices helped shape the new boundary, system builder remains necessary. Table 1 lists some of the main interests and the different actors.

A designer is an integrator and a negotiator; hence, Saarinen's success as a system builder is the product of posing the problem in a way that any possible outcome would cover most of the interests on the negotiation board. In spite of the excessive reliance of FAA in one person, Saarinen came afloat, projecting confidence and reliability. His use of a scientific approach to his designs granted him invulnerability against the enemies of change. Much was at stake in this endeavor for both the designer and the promoter; faith on the efficacy on a new landside–airside boundary relationship was the only way to guarantee a welcome reception to the reinvented airport.

As the landside–airside boundaries gained complexity, they could now be materialized or de-materialized, meaning that they could be designed to shortening the total passenger's walking distances. In the case of Dulles terminal, the concourse level may be fully considered part of the landside. Nothing at the building is subordinated to the interests of the airport, not even the height of the floor. Conversely, no aspect of the aircraft enforces actions on the part of humans. At Dulles, the landside–airside boundary materializes as an independent artifact, adding even more complexity to its historical relationship. The Mobile Lounge is like being in limbo or in a shared territory of convergence. If passengers are neither in one realm or the other, the Mobile Lounge reveals as a nonplace that is far from neutral. For the first time in history, a "bustronaut"[16] (as drivers used to call themselves) was in charge of adjusting the landside–airside device and make it "fit" according to the needs of

[16]Sutton (1962) mentions in his article that the Mobile Lounge drivers used to call themselves "Bustronauts" (p. 23).

Table 1 Actors who intervened in the decision process for Dulles, by main interest and specific role during negotiations

Social group	Central interest	Specific role
FAA (Federal Aviation Authority)	Program	Provides a new kind of airport that responds to the current airport crisis across the country. The agency showed preference for a variation on the UK's bus system
The main contractor	Deliver	Delivers a master plan that is technically sound and complies with the contract
The designer	Integrate	Creates a paradigmatic building that stands out in architectural modernism and is therefore accepted by the leading groups of the profession. Outshines his competition (St. Louis Airport)
The designer's team	Justify	Technically justifies the new solution, provides evidence (statistical and quantitative) in the shape of "exhibits"
Chrysler		Engineer that technically solves the Mobile Lounge. They bet for a potential airport dependency in the short future (ML devices) and thus future business
Users of airports	Demand	Want a more comfortable experience of travel; complain about walking long distances or multiple levels, agglomeration or jamming, getting exposed to the elements, etc.
Critics and journalists	Push On	Have become ironic and sour. Complain about the airports of the 1950s is so cramped and uncomfortable, that is becoming the news. Places hope on new ideas that could improve the current situation

(continued)

Table 1 (continued)

Social group	Central interest	Specific role
The airlines	Trade	Did not like the concept of Mobile Lounges because it would create added cost and additional need for personnel, thus, less revenue. Refused to take any responsibility for the "experimental" layout
The administrators	Decide	Agreed upon the need to look for options other than the "finger scheme"
ICAO (International Civil Aviation Organization)	Regulate	Issued regulations that expected designers to address the landside–airside boundary in their designs. The norm called for multiple requirements such as separation of flows
Pilots and aircraft owners	Improve	Asked for equal convenience at the terminal building for seat-holding airline patrons; no discrimination for those arriving in transportation other than cars, equal dealing with users of general aviation
Federal government	Foresee	"Super-airports" based on exceptional planning and located at a considerable distance from metropolitan centers. High costs could be anticipated
The airport consultant	Prioritize	The specialized voice of this specialist has been deliberately suppressed by the client. Airport consultation could end in a "priorities conflict." Rather, the FAA takes all the credit for what they called "airport studies"

machines and humans. On the one hand, these "bustronauts" would have to manually operate the machine and adjust height in order to flush the terminal (as a passenger airbridge); on the other hand, the driver changes the machine configuration and adapts it to the variable height and specs of aircraft. In this regard, this is a fascinating case of technological mediation and possibly a unique study of human vs non-human actorship.

But perhaps only Quesada's clairvoyant view describes why this discussion is right at the core of technological determinism. In 1958, the President's Airport Commission maintained that "the question whether fixed airport facilities should control future airport design is highly controversial. To say that the average-capacity should be the limiting factor in airport design is to place an undesirable inhibition on progress. Conversely, complete latitude in airport design planning makes long range planning for a well-integrated airport system difficult or impossible" (President's Airport Commission, 1958, pp. 38, 39). This debate captures how the opposed interests of technical groups and users will eventually collide and form landside–airside boundaries. On the one hand, new aircraft "demands" new requirements; for example, a 1958 DC-8 jet by Douglas was significantly larger in wingspan, and higher and noisier than the previous DC-3. On the other hand, passengers no longer wished to be exposed to the high-pitched noise from the jet turbine; for the traveler walking on the tarmac, it was not as romantic as it used to be. How should designers solve this conflict?

I suggest that landside–airside boundaries are *technological interfaces* with two ends: at each end, humans and machines interplay, in a continuum state of negotiation. That is why the airport never stops changing. But once identified as an entity (an interface in this case), it can be studied and classified. Just as the interface of a device mediates the interaction between users and technological artifacts, airports would become the interaction platform between users and aircraft.

Ultimately, a landside–airside boundary works as a technological anchor that allows the understanding of the system from top to bottom, and not necessarily from the base up. The way this operates may be exemplified by the Dulles concept. For example, I described earlier in this chapter how different airport layouts failed to provide a better balance to the human (user)/machine (transport) conflict, and I pointed out how they formed a sort of technological progression, up to the point where innovation, or the jet-bridge, becomes an evident solution to the crisis. However, the outcome of Dulles is totally unpredictable in terms of historical linearity. The FAA showed no "*blind faith in progress*" because the preferred option for the airport was not the obvious pier plus jet-bridge. In this situation, the futurologist cannot predict what is next, and the analyst no longer understands how the system works anymore. On the contrary, the sole identification of the landside–airside boundary at Dulles might explain why such a layout addresses almost twenty years of planning failures. Each question that we ask about the way Dulles's boundary was set will inevitably unveil an old chain of techno-social conflicts and an intriguing historical irregularity.

Under a cloud of great expectations, the FAA reported that the "new Dulles International Airport was being constructed with passenger comfort and convenience in mind. For instance, instead of walking, passengers will ride in air-conditioned comfort from the terminal building to their airplanes in a waiting room on wheels. Called a Mobile Lounge, this unique vehicle will be introduced to the air traveling public at Dulles International Airport when it opens for business in the fall of 1962. In size it is comparable to eight intercity busses arranged with two side by side, two more parked to the rear of these and then four more stacked on top of the first four" (FAA, 1961a, p. 1). According to the *Birmingham Eccentric,* more than "30,000 of critical mock-up situations were tested before the opening" ("Mobile lounge-ramp ends walk to planes," 1962). But how was the new airport received by the public?

In the inaugural dedication speech, President John F. Kennedy stated, "For years, people have been hearing about a revolutionary airport for Washington!" (FAA, 1962). The airport kicked off with 566 flights in the first three weeks of operation (Bacon, 1962a). "It's open now, and questions rise: What is it like? Will all Washington flights land there? Are novel features working well?" (If you're flying to Washington, 1962, p. 82). The *St. Louis Globe* (1962) masthead read: "Dulles National Airport makes hit with travelers." According to letters, during the first week of operations, passengers complained about the inability to get off and use and enjoy the terminal—the Mobile Lounge concept forced all passengers in transit to remain on board until staying passengers deplaned.[17] *Travel Notes* observed that "the Mobile Lounge eliminates the cost of building fingers for those long cardiac causeways that lead out to the jets in many airports, here and abroad" (Sutton, 1962). For the *Boston Globe,* "the Mobile Lounge is FAA's answer to the problem most major airport are faced with today—that of the ever growing distances passengers are required to walk between the terminal building and their aircraft. Up to 90 persons will be transported in the Mobile Lounge. It is the largest passengers carrying vehicle ever built to be operated on rubber tires" (Riley, 1962). The *New Haven Register* titled a column "The People Likes It: New Dulles Airport Starts Out as $110 Million Hit" (Haugland, 1962), and Sutton (1962) declared that at the Terminal "no one need touch a tootsie to the crust of the outside world. No steps, no stairs, no slush, no slop. The Mobile Lounge has removed the last mile from air travel, at least in Washington" (p. 23). For the *Boston Globe,* it was "designed to take the walk out of airports, it has been hailed as the greatest thing for the human foot since the advent of arch-support shoes" ("Logan still a winner," 1965). Even the combative association of pilots and aircraft owners praised that in Dulles, "both designers and planners for Washington's 9,800- acre jet airport have been doing advanced thinking about facilities to increase safety and efficiency in the handling of international flights. Today's jet airport, frequently a monument to national or civic pride, can be managed to spread the cost load by designing the landing areas and terminal building to accommodate a substantial increase in that general aviation traffic which complements the airline business" (Thompson, 1962, p. 1). For the association, "the nation will be proud of Dulles airport as a national monument and service facility if its final design and operations plans are also realistic in providing for domestic long-haul patrons" (Thompson, 1962, p. 1).

As an outcome of this story, the new landside–airside boundary turned quickly into a cultural icon. By the end of 1962, *My Weekly Reader*, a periodical also called the "Children's Newspaper" devoted its November issue to airports, with a focus on the newly opened Dulles

---

[17]Bacon, R. F. [Director of the Bureau of National Capital Airports]. (1963, May). [Letter to the Administrator of the FAA].

("A Jet Airport Opens," 1962; see www.landsideairside.com). In the article, many references are made to the landside–airside boundary in a compelling, simple way article explains in ordinary language how the airport borderlines are organized in different levels; it describes the flows, the waiting, and the new conveniences. It also includes a short reading comprehension exercise in which children are asked questions that make them think of airports and the new features of the airport.

The cultural dimension of these expressions offers an invaluable opportunity to understand the social reverberations of the landside–airside boundary and the airport. No less compelling was the fact that Mobile Lounge drivers felt proud of their historical task. The term "Bustronauts" wonderfully captures how Dulles and its shiny new landside–airside boundary were understood as the future here and today for those traveling through or working on its grounds.

The FAA spoke of the future expansions with great optimism: "By 1975, when additional gates have been constructed, some aircraft will be parked as *remote positions* as far as three miles from the terminal building." But these dreams never came true. The Mobile Lounge concept lasted for a few years, but the introduction of security measures through ICAO's *Annex 17* conflicted much with the rigidity of the scheme. By the early 1980s, Dulles Airport's main tenant, United Airways, decided to build midfield buildings where there should be only aircraft parking per Saarinen's layout. Those temporary buildings were necessary to create more hold-room capacity and to release pressure on the declining Mobile Lounge. During my last study visit to Dulles in 2010, the Mobile Lounge was working as an inter-terminal shuttle, and boards on the walls announce its final replacement and removal for good. Perhaps in the future Dulles will be to airports what the Concorde is to aviation.

In conclusion, I have used Thomas Hughes's metaphor of technological momentum to describe how it led to the (somehow frustrated, however) development of a super-system of airports. I have analyzed how Dulles, just like LaGuardia, may be considered a historical *irregularity*. Thus, in Hughes's terms, Dulles may be understood as a "reverse salient" protruding out of an imaginary timeline of "standard" airport terminal designs. In order to reduce the nightmares of delay and the rising popular discomfort in airports, the FAA initiative of reinventing a new kind of airport prompted the development of new ways to deal with the rising landside–airside boundary conflict. Designers were encouraged to rethink the airport as a socio-technical system, and this incentive shaped the experimental planning behind the new airport. In this section, I recount how faith in technology at the turn of 1960 contributed to the idea that the Mobile Lounge was the ultimate solution for the modern airport and its modern user. However, during the following decades, Dulles's influence reached the far corners of the world, but paradoxically the airport itself was stigmatized as a failed idea. Flashy, young, professionalized airport planners harshly criticized the new system and wish to propose a new one. In the next chapter, I will tell their story.

References

Publications

Activities are channelized and separated: Washington National Airport. (1941). *Architectural Record, 90*(1), 53.

Airport bus would haul passengers right to plane. (1953, July). *Popular Science.*

The airport scramble. (1956, June). *Architectural Forum.*

Akrich, M. (1992). The de-scription of technical objects. In W. Bjiker & J. Law (Eds.), *Shaping technologies*. Cambridge: MIT Press.

Ammann & Whitney, Eero Saarinen, Burns & McDonnell, Ellery Husted, & Burhan Kelly. (1960, May). *Master development plan report for Dulles International Airport.*

Arroyo, N. R. et al. (1953, May). Airport terminal buildings: Twentieth century building type. *Progressive Architecture.*

Bacon, R. F. [Director of the Bureau of National Capital Airports]. (1962a, March). [Letter to U.S. Congressman M. Gerhardt].

Bacon, R. F. [Director of the Bureau of National Capital Airports]. (1962b, April). [Letter to the Administrator of the FAA Memorandum No. 7941]. From the U.S. State Government.

Bacon, R. F. [Director of the Bureau of National Capital Airports]. (1963, May). [Letter to the Administrator of the FAA].

Bednarek, J. R. D. (2001). *America's airports: Airfield development, 1918–1947*. College Station, TX: Texas A&M University Press.
Bijker, W. E. (1997). *On bicycles, bakelites and bulbs: Toward a theory of socio-technical change*. Cambridge, MA, and London, UK: MIT Press.
Billington, D. (1983). *The tower and the bridge: The new art of structural art*. Princeton, NJ: Princeton University Press.
Bird, A. (2011). Thomas Kuhn and the concept of paradigm. In E. N. Zalta (Ed.), *The Stanford encyclopedia of philosophy*. Stanford, CA: Center for the Study of Language and Information.
Boyer, P. (1985). *By the bomb's early light: American thought and culture at the dawn of the atomic age*. Chapel Hill, NC, and London, UK: University of North Carolina Press.
Chandos, J. (1956). Passenger progress. In *London Airport: The official story of the new world air centre*. London, UK: Ministry of Transport and Civil Aviation.
Downs, T. (2009, May 13). An introduction to airport design. *Dwell*.
Dunbar, B. (1960, March). Airport breakthrough or bust… FAA's Dulles plan is different. *Airlift*.
Federal Aviation Agency. (1961a). *Dulles International Airport: Fact sheet*.
Federal Aviation Agency. (1961b). *FAA's Office of Public Affairs Newsletter*. Washington, DC: Author.
Federal Aviation Agency. (1961c). The mobile lounge: Fact sheet. *FAA News*. Washington, DC: Office of Public Affairs.
Federal Aviation Agency. (1962, November 17–18). *Dulles International Airport: Dedication program*.
Froesch, C., & Prokosch, W. (1946). *Airport planning*. New York, NY: Wiley.
Geraci, P. (1965, June). World's safest airport. *Rotarian*.
Haugland, V. K. (1962, December). The people likes it: New Dulles Airport starts out as $110 million hit. *New Haven Register*.
Hughes, T. P. (1983). *Networks of power: Electrification in western society 1880–1930*. Baltimore, MD and London, UK: Johns Hopkins University Press.
If you're flying to Washington. (1962). *U.S. News & World Report*.
International Civil Aviation Organization. (1953). *Aerodromes: Annex 14 to the Convention on International Civil Aviation* (2nd ed.). *International Standard and Recommended Practices*. Montreal, QC: Author.
Jet-age airport. (1959, December). Bonanza from Billy: Art. *New Yorker*.
Jet-age progress. (1959, April 12). *New York Post*.
A jet airport opens. (1962, November 26–30). *My Weekly Reader: The Children's Newspaper*.

Johnson, P. (1946). Postwar planning reconversion. *Aviation Digest Index*. Washington, DC: U.S. National Research Bureau (Journal).
Karsner, D. (1993). *"Leaving on a Jet Plane": Commercial aviation, airports and post-industrial American society, 1933–1970* (Doctoral dissertation). Temple University, Philadelphia, PA.
Lackney, J. A. (1999, September). *Assessing school facilities for learning/assessing the impact of the physical environment on the educational process*. Mississippi State, Miss.: Educational Design Institute.
Loading made easier. (1952). *Flight*.
Logan still a winner in time. (1965). *Boston Globe*.
A look at the newest thing in airports. (1962, September 10). *U.S. News & World Report*.
Mobile lounge-ramp ends walk to planes. (1962). *Birmingham eccentric*.
Mumford, L. (1934). *Technics & civilization*. Chicago, IL: University of Chicago Press.
New thinking of airport terminals. (1956, June). *Architectural Forum*.
Planning of jet airports, passenger terminal building design principles. (1960, March). *Architectural Record*.
Projected Airport by Fellheimer & Wagner Architects, L.L. Odell Aviation Consultant. (1940, August). *Architectural Forum*, pp. 85–87.
Prokosch, W. (1951, January). Airport design: Its architectural aspects (Architectural Record's Building Types Study, Number 170). Architectural Record.
Riley, A. A. (1962, November 25). At last—An airport that's easy on the feet. *Science in Industry,*.
Rosenberg, N. (1994). *Exploring the black box: Technology, economics and history*. Cambridge, UK and New York: Cambridge University Press.
Ruark, R. C. (1961, May 2). The airport dilemma. *New York World Telegram*.
Scott, J. C. (1998). *Seeing like a state: How certain schemes to improve the human condition have failed*. New Haven, CT and London, UK: Yale University Press.
Smith, Hinchman & Grylls, Inc., Architects and Engineers. (1947, October). Establishing an airport planning program. *Architectural Record*, pp. 90–94.
Sutton, H. (1962). New job: Bustronaut. *Travel Notes*.
Szurovy, G. (2003). *The American airport*. St. Paul, MN: MBI.
Temko, A. (1962). *Eero Saarinen*. New York, NY: George Braziller.
Thompson, C. (1962, September 1). *The AOPA airport letter*. Washington, DC: Aircraft Owners & Pilots Association.

Time study—Mobile lounge, Kodak color patch. (1959). *Saarinen Archive.* Manuscripts and Archives, Yale University Library.

Unique mobile lounges: Dulles National Airport makes hit with travelers. (1962, December 10). *St.Louis Globe–Democrat.*

U.S. Civil Aeronautics Administration. (1953, April). *Airport terminal buildings.* Washington, DC: U.S. Department of Commerce.

The U.S. President's Airport Commission. (1958). *The airport and its neighbors.* Report of the U.S. President's Airport Commission. Washington, DC: Government Printing Office.

Vincenti, W. G. (1976). *What engineers know and how they know it: Analytical studies of aeronautical history.* Baltimore, MD and London, UK: Johns Hopkins University Press.

Washington National Airport. (1941, October). *Architectural Record, 90*(4).

Work Projects Administration. (1938). *A detailed description of the great air terminal.* New York Municipal Airport. I.T.S. Library University of California at Berkeley.

You ride to plane in mobile lounge. (1962, November). *Sunday Star.* Washington, DC.

4

"The Landside Airside Concept": Breaking to Reconnect—The "People Mover" at Tampa International Airport, 1962–1971

Right after Dulles Airport was inaugurated in 1962, a group of young professional airport planners (Leigh Fisher and Associates) started working once again on a complete reinvention of the airport. In a house's basement, they came up with the "Landside Airside Concept," or a new, revolutionary layout that would surpass both the experimental Mobile Lounge at the capital's airport and all other previous schemes.

This chapter tells this story in six sections, which mainly spin around the voices of those actors who were involved in the planning of the fascinating airport that is Tampa International Airport (TIA). In the Sect. 1, I talk about the way I intend to apply S&TS concepts and introduce the reader to the general framing of the analysis. In the Sect. 2, I discuss the history of Tampa's airport and identify the actors who took part in this story (particularly in the formation of the landside–airside boundary), and present a chronicle of the public reception of the new facility. Although this may be anticlimactic in narrative terms, I would like to provide first a full historical snapshot of the airport planning.

In the following four sections, I analyze the interactions of the design's stakeholders and the role of planners. I have arranged these discussions around three key issues: "problem posing," "problem construct," and "problem solving." I have also framed the discussion in

V. Marquez, *Landside | Airside*, https://doi.org/10.1007/978-981-13-3362-0_4

technical terms, using Wittgenstein's (1970) language interpretation in *Zettel*. Thus, in Sect. 3, I speak of Leigh Fisher's diagrammatic and semantic planning approach employed at Tampa, directly from interviewing those involved at the time. Next, in Sect. 4, I address the technical procedures and methods, or what I call the problem construct; in Sect. 5, I deal with the problem-solving, or in other words the final planning process. Last, in Sect. 6, I conclude by pointing out what happened after Tampa's airport in terms of planning and the consequences of extending the People Mover into other airports. In the end, I discuss why airports are open negotiations that will hardly find a technological closure.

1 Framing Tampa

The analyst may be tempted to assume that after the success of Dulles, most airports followed the adoption of the new and ostentatiously efficient system of organization, including Mobile Lounges. But surprisingly this was not the case. Most large international airports—among them Chicago, London Heathrow, and Miami—kept growing in the same old fashion, expanding on lengthy piers and concourses either to the sides or perpendicularly to the "box" terminals. In spite of the growing passenger discomfort and harsh criticisms, airport authorities paradoxically chose the perhaps safer "standard" airport layout.

Although Dulles proved to solve the biggest problem of the modern airport by reducing the walking distance from drop-off to enplaning to a minimum, the new technological artifact did not reach closure, and some of the relevant social groups, particularly the airlines, remained distrustful and unconvinced. However, physical expansion in response to the growing traffic demand was inevitable, and standard designs remained as a pattern of stabilization.

Of the three noticeable "reverse salients" (Hughes) in what I call in this book a *linear history* of airport planning, I have addressed just two so far. In Chapter 2, I described how New York LaGuardia Airport was a techno-political artifact in which the interests of politicians finally imprinted the materiality of the design (Skywalk). In Chapter 3, I showed how Dulles International Airport captured the ambition of the designer and the state powers who endorsed him (Mobile Lounge). In this chapter, I present how TIA brought the first opportunity for airport planners to rethink an airport from the ground up (Landside–Airside Concept).

As the analyst could easily predict, most of these practices have been jealously black-boxed by airport planning companies through the years. Thus, I must thank in advance the generosity of the interviewees, because their voices and experiences are the real backbone of this chapter. More specifically, this chapter explores in depth the heuristics behind the landside–airside boundary and responds to why this problem-solving activity led airport planners to reinvent the airport once again. In this sense, and as a research tool, I am using Bijker's theoretical concept of the technological frame in order to explain why these processes comprise social, material, and cognitive elements. Bijker proposes a tentative list of elements comprising a technological frame. These elements could "structure the interactions among the actors of a relevant social group" (Bijker, 1997, Ch. 3). In the case of Tampa's airport planning process, the technological frame may be simplified as shown in Table 1.

In this chapter, I discuss changes in the landside–airside boundary as a result of the world's massive growth of air travel as well as the new availability of planning experts. Beyond pressures from concessionaries and the ever-changing technological conditions of aircraft, the fact that airport planning became a well-established field of specialty in universities and technological institutes had an enormous impact in the way airports were conceived. The paramount example is TIA of 1971, in which a young breed of planners pioneered the way in which the landside–airside boundaries could be materialized in a different spatial dimension called the "Landside Airside Concept" (1963) (see Fig. 1). In this case, one of the leading airport planning consultants in the country, Leigh Fisher and Associates (LF&A), and not the FAA, would persuade the officials of Hillsborough County to re-study the quantifiable inefficiency of current airport terminals. The outcome was a radical innovation, which became an instant airport planning touchstone and a model that influenced new airports worldwide. The landside was

Table 1 How Tampa International Airport Meets Bijker's criteria and how it may be analyzed as a technological frame

Bijker's list	Features of case study (Tampa)
Key problems	Users' discomfort, lengthy walks, congestion; aircraft change moves faster than airports adjustments
Diagnostic	Historical contradiction between opposite land-side and airside demands
Semantics/communication	Definitions: landside, airside, terminal, purgatory, concourse, people mover, etc
Requirements (negotiated)	750 ft. maximum tolerable walk from drop-off to enplanement (International Air Transport Association)
Methods	Comparative, statistical. Between 20 major U.S. airports
Current standards and styles	Linear or perpendicular terminal growth (piers); enfield terminals
Hypothesis, solutions	Landside–airside split; midfield terminal
Constraints	Reconnection: implies new technologies. High cost of innovation
Discourse	An airport for people and not for machines

physically separated from the airside, but linked through a "People Mover," shortening the long walks and avoiding the so-called delay. For the first time in more than three decades, zones were again called "airside" and "landside," and the airport was now considered a transportation hub (or a multimodal facility, in contemporary terms). In fact, the airport planning company rapidly copyrighted the report describing the scheme. Through a series of interviews, this section frames the interactions of those actors who negotiated the landside–airside boundary during the planning phases at TIA.

2 The Airport and Its Actors

"I was lucky to get here at the very beginning," Bean said. "The concept of the main building being separated from where people get on and off airplanes— connected by some sort of ride system—was totally revolutionary. I knew very little about Tampa, but I was told it was in the

Fig. 1 A view of the centralized "landside" building and its "airside" satellites. Notice the interconnecting piers bridging the hard frontier between apron and road. *Aerial view of Tampa International Airport* [digital] (1972). Public domain (Provided for reproduction by the Florida photographic collection at the Florida State Library)

process of doing something revolutionary." I came down and said, "This is where the future is." (as quoted in Waresh, 1987)

Tracing the main actors who participated in landside–airside negotiations may be difficult. In this sense, I follow the structure used by historian Deborah Douglas (1996), in her keystone dissertation *The Invention of Airports: A Political, Economic and Technological History of Airports in the United States: 1919–1939,* in which she claims that the evolution of airports required the collaboration of architects, engineers, city planners, and local and federal authorities. Leaving aside a few isolated cases such as the visionary airport consultant J. J. Odell, who started consulting in the 1930s, the birth of airport planning would not be consolidated until the end of the Second World War. During those days, pioneer firms started airport consultation to provide financing, planning, and engineering studies. Among these firms, Buckley, Landrum & Brown (L&B) and Leigh Fisher and Associates (LF&A) stood at the forefront. L&B was characterized by its technical "rigidity" and low "negotiation" capacity; it was a well-balanced firm equipped with architects, engineers, and financiers. LF&A was founded by Leigh Fisher, an engineer, who has been described as a "strong negotiator" and knew how to surround himself with the best in the industry (Brink Coridan, 2006, p. 2). Even in his early years, the firm was a sound success, handling a client portfolio of more than 90, and had worked by 1959 on more than 140 civil airport projects.

But besides the establishment of airport consultation as a field of specialty (which was mostly learned from experience, by the way), other factors helped trigger strong changes in the way airports were designed in the decade to come: "In the late 1960s, air traffic had been exploding across the nation and the world. In response to burgeoning air traffic demand, terminal buildings throughout the nation were being expanded by adding new concourses and lengthening existing concourses. As a result, walking distances at major airports were becoming excessive and the media were almost daily criticizing the 'heart attack' walks at most major airports. At Drew Field, air traffic demand had far exceeded the capacity of its small 'box' terminal to accommodate it" (Brink Coridan, 2006, p. 2). This situation triggered an alarming congestion in the small terminal by the early 1950s: "We are facing the prospect of an aerial traffic jam that could slow down the state's business growth" ("State's Skies Getting Congested," 1989), declared Hillsborough County—before starting the analysis for replacing the existent airport—and picking through a long, heated debate, between St. Petersburg and Tampa (see Bliss, 2011) as the new potential location for the new airport.

According to the Hillsborough County Aviation Authority (HCAA, 1965), "at peak hours, queue lines at the airline ticket counters sometimes stretched across the airport entrance road." It was clear that a new terminal was desperately needed. Rather than simply go full speed ahead with a conventional terminal concept, Leigh Fisher, who had been consultant to the HCAA since 1957, convinced the Authority to undertake a research study of existing airport terminals. This study would be directed toward evaluating passenger comfort and convenience, which had become the focus of sharp newspaper and media criticism. The intent of the study was to identify passenger service problems and, in particular, to quantify walking distances at major airports as a basis for planning new terminal facilities at Tampa. The real emphasis, as it turned out, was on lengthy walking distances (because they could be quantified) and on possible solutions to the walking distance problem. Fisher led the research team, which started with a major research study of 20 US airport terminals; the study was overseen by Jim Jones and prepared by the young Marjorie (Marge) Brink Coridan, the first female airport planner worldwide. In her corporate memoirs, Brink Coridan (2006) recalls those early steps in the process[1]:

> The study concluded with alternative planning concepts designed to solve the walking distance problem, some prepared by Jim Harrison and some by Gene Lewis, who was retained independently by Fisher at that time. The result of the <u>research and planning studies was the "landside/airside" separation concept</u>, which separates the permanent functions of the landside (ticketing and baggage claim) from the continually changing functions of the airside (changing aircraft designs and operations) and connects them with a ride system. It is a source of minor pride that <u>the terms "airside" and "landside," standard today in the industry lexicon, were born in Jonesy's basement during a review session of the research report</u>. It is also a source of minor pride that the research report is the only report ever copyrighted by LF&A. (p. 57)

These last two sentences are particularly interesting and will be addressed in the following section, but the tone of the quotation is worth highlighting here. It is hardly deniable that these experts truly felt that they were in front of an imaginary device that was powerful in terms of both problem posing and problem solving. Brink Coridan (2006) continues:

[1]In the following citations, I have added emphasis, underlining the words, phrases, or passages that help to trace the origins of their Landside–Airside Concept, as well as their own opinions about the landside–airside boundary in general.

> As a result of the research and planning studies, it was recommended that the Authority (Hillsborough County Aviation Authority or HCAA) explore a new terminal design concept in which the landside and airside functions were separated and connected by a ride system. When the program was initiated, Herbert C. Godfrey, Jr., was Executive Director of the Authority. After the adoption by The Authority (HCAA) of the recommendation to pursue the landside–airside separation concept in parallel with a conventional terminal concept, Godfrey decided to resign from the Authority to assume a vice presidency with United Airlines. With Godfrey's departure, the then-Deputy Executive Director, George J. Bean, assumed the position of Executive Director. There could not have been a better successor. Bean immediately formed a design team and a management committee to oversee the planning and development of the new terminal facilities—and established a calendar- based meeting schedule to ensure that the program kept moving ahead. The team included Leigh Fisher, Leigh Fisher Associates (Bob Joerger); Allen, Dell, Frank, and Trinkle, attorneys (Stewart C. Eggert); Greiner Engineering (Rene Crouch), and two architects, initially, Carlos Alfonso, a talented local architect with a small company, who was succeeded by Reynolds Smith & Hills, a major architectural firm which had the depth of personnel to handle a project of this magnitude. (pp. 57–58)

Here, Brink Coridan leaves clear the chain of command, which obviously has a strong impact in assigning the right *decision weight* of each actor: the administrator, the planner, the lawyers, the engineers, and the designers. It is striking to learn how, compared with my previous examples, in which designers like Delano and Saarinen played protagonist roles, in this case, Brink Coridan plays a support role within a larger cast. This is especially interesting when we think of the magnitude of reinventing a technological system like the airport; the "system builder" finally made its way to a new, more complex negotiation process. Brink Coridan (2006) closes her summary:

> In the end, after six years of intensive planning and design efforts, the new terminal complex, complete with a "must ride" Westinghouse System, opened in April 1971 to international acclaim. The Westinghouse System, upon which the operation of the terminal complex was completely dependent, had never been tested in a formal

> application before. In addition to the unique and innovative design concept, the management aspects of the program were equally creative. In order to ensure the retirement of the revenue bonds (first issue: $67 million), the Authority adopted a formal concession policy designed to maximize revenues while providing the highest level of customer service.
>
> This policy was adopted as a condition of the revenue bond indenture. During the concession award process, various adjustments were made in the pre-bid and pre-proposal conferences in order to ensure that a certain level of revenues was guaranteed while permitting the concessionaires to amortize their investments. For example, in the news and general merchandise agreements, originally intended as five-year agreements, the concessionaires agreed to take significant losses (called a "blood bath" internally) in the first five years of the agreement and make up their profits in the last five years. The 10-year Authority audit reports showed that this was exactly how it worked out. The Tampa International Airport terminal facilities and management program have been a model in the industry. When LF&A became the consultant to the Greater Orlando Aviation Authority for the development of its new terminal facilities, John Wyckoff, the Authority Deputy Director and manager of the Orlando terminal development program, instructed LF&A to "do exactly what you did at Tampa." (p. 59)

In the 1960s, the manager of Atlanta Hartsfield International Airport, George Bean, attributed the planning and design concepts that had been introduced successfully in Tampa with inspiring the innovative concept for the new midfield terminal complex in Atlanta, which opened in 1981. Bean was a legendary airport director. LF&A was privileged to serve as the HCAA consultants from 1957 until Bean's retirement in 1996. Bean, ever gracious and ever generous, always gave credit for the success of the new Tampa terminal facilities to Leigh Fisher. Beyond reflecting the LF&A view of this long enterprise, this vivid account represents the voice of the airport's main planner and also serves as a candid introduction to the TIA planning route.

The old Drew Field was a hard element to replace, however. This was a very popular airfield in its time, a military facility turned into commercial airport and equipped like no other facility of its kind in terms of leisure. Perhaps only Pittsburgh matched the array of entertainment options. Among others, it was furnished with a pool and theaters which showed films "weeks ahead of regular Theater circuits" (Hawes, 1978). But throughout the years, it grew obsolete and became incapable of handling the booming traffic. Surprisingly, the new building complex was far from aesthetically pleasant; it was miles away from the grandiose rail terminals of the past and the *cathedralesque* modern masterpieces such as TWA. "It's not designed for beauty, but new terminal is impressive" ("New 180-foot control tower due," 1971). According to the *Tampa Tribune,* Authority director Bean said that the HCAA "did not want the building to be a monument, as a lot of airports have become"; "because the building is so simple looking—it's imposing because of its size—it ended up not being as attractive as we would have liked it. So, to increase the beauty of the complex, we decided to use a great deal of landscaping" ("Greening of an airport," 1971).

The press always showed an enormous interest in the airport. Several magazines and periodicals described even the planning process as something worth speaking of, with a certain sense of pride. Author Michael Bane wrote in *The Best Li'l Ole Airport* (1981) that

> the authority hired aviation planning consultants Leigh Fisher Associates of San Francisco and instructed them to spend the next six months just looking at other airports analyzing their collective strengths and weaknesses, and then to develop something better. What the survey found was that when people came to airports, the thing they hated most of all was long walks to the planes. Most air terminals had been built for convenient parking, and as the airplanes got bigger and the number of flights increased, "fingers" stretched away from the main terminal to the planes, and the passengers had longer and longer walks, the people-moving system, which whisked 100 passengers back and forth between Airside and Landside in less than a minute, was so popular that the HCAA "opened" the airport a few weeks before the official opening to allow local residents to ride back and forth in the shuttles, which proved to be an immensely popular pastime. As the Chamber of Commerce is fond of pointing out, Tampa International Airport has been singled out by such publications as *Fortune* and *Esquire* as one of the best—if not the best airport in the country. (Bane, 1981, pp. 104–106)

I find it particularly intriguing that those who were commenting and publishing on the airport were not trying to translate or interpreting the terms landside and airside, but the contrary. Most publications refer with great familiarity to both concepts without offering much explanation.

Journalists and users rapidly appreciated the new features of the airport. The *Tampa Tribune* article "The New Air Terminal: A Site for Sore Feet" noted, "In an industry focused on engineering and high technology, Tampa International Airport is generally known as the first airport designed for people, not machines" (Garret, 1976). In a published interview (Garret, 1976), Bean, the director of the HCAA and the "man behind" the realization of Tampa, explained that "crowded airport terminals trying to accommodate increased flights and larger aircraft attempted to solve their problems by constructing fingers or piers. This resulted in longer passenger walking distances, which was totally unacceptable. Dulles Airport (outside Washington, DC) attempted the first innovative busing of passengers to airplanes from the main terminal. Our design team studied their concept and improved it. The airport was designed for people comfort and convenience rather than specifically for airplane convenience." Bean was quick to add that Tampa International maintained a close working relationship with the airlines. The airport's design had attracted worldwide interest, with visits by 70 foreign delegations from Afghanistan to Zaire and Zambia to study the airport facilities.

In 1975, Paul MacAlester, the former director of information of the HCAA (as quoted in Garret, 1976), described how "back in the 60s when this *daring concept* was being formulated, Hillsborough County Aviation Authority board members vigorously recruited support of far-sighted business and political leaders in Tampa. Although the Authority is an autonomous unit of county government and vested with the full prerogatives of a private corporation, MacAlester said the project still needed community support. The authority owes a special debt of gratitude to its own former chairman, Clyde Perry, who spearheaded the endeavor; to Leigh Fisher Associates, charged with planning, and Herb Godfrey, former HCAA director, who helped the design team come up with the airside/landside terminal concept." MacAlester had reasons to think the concept was *daring*. According to him, the airline officials were strongly skeptical of the scheme, and it was thanks to the negotiations by two members of the HCAA, Norm Stallings and Sam Bucklew, that the acceptance of airlines was finally gained. "It was a painful project to see through to completion," recorded Tom Inglis, associate editor of the former *Tampa Times*. "There were grand jury investigations, special investigations at the orders of the Governor, lawsuits, strikes, and other problems" (as quoted in Garret, 1976).

The airport at Tampa was one of the first facilities built right at the transition to the enforcement of the first steps in civil security. The American best-selling novel *Airport*, by Arthur Hailey (1968), and later the more popular and iconic film *Airport*[2] (Seaton, 1970) exploited for the first time the potential fears at the airport. The plot refers to harsh weather, congestion, and airport management as potential threats. More disturbingly, it presents, in an almost naive way, the case of a desperate man, D. O. Guerrero, the first suicide bomber in history. Audiences panicked. Hence, Tampa was a pioneer airport in setting up security measures. Airport Police Major J. Quinn declared to the *Tribune* that "the airport has completely adequate emergency facilities of its own, of course, Quinn adds. And Tampa International has had emergencies. 'Sure, we've had 'em.' In addition 'especially since hijacking became popular,' the airport has had potential security problems to prevent. It also has traffic problems, lost or confused patrons, bomb threats and big areas vulnerable to vandals. Because of all this, Tampa International had its own police force, almost militarily detailed emergency procedure plans, and special emergency and security equipment" ("Airport police prepared," 1971. p. 4).

Another important aspect, key to the discussions around landside and airside, is that aircraft innovation presents a continuum of changing demands and requirements. As this pattern is unpredictable, each facility must be adapted in one way or another to service the most current equipment. These design features are very often a source of pride

[2]Seaton, G. (Director). (1970). *Airport* [Film]. Los Angeles: Universal.

among planners and critics. "The huge new DC-10 is one of the jumbo jets for which the Tampa International Airport was designed and constructed. Along with the already heralded Boeing 747 and the Lockheed L-1011, it will be seen regularly landing at the new airport, the first in the nation designed specifically with the jumbo jets in mind," wrote a journalist with the *Tampa Tribune* ("DC-10 will fit nicely," 1971). A Northwest official assured that "all gates are equipped to handle the 747 and DC-10 aircraft in addition to other planes in Northwest's fleet" ("Those faraway places," 1971).

Another unusual feature is the location of the lodging facility. "The Tampa Airport Marriot Hotel, located as the centerpiece in the magnificent Tampa International Airport, is one of the assets that make the Tampa airport a monument of convenience in air transportation" (Dunn, 1985). For journalists, Tampa was an abundant source of reportage. The *Times*, describing the new system, said that "the airport's unique system of speeding travelers from the main building to the satellite or 'airside' buildings is usually referred to as a 'people mover' apparatus" ("Getting there's easy," 1971). Howard Gorham (1971c), another writer with the *Tribune,* chronicled in his article "They Came, They Saw, They Liked" how the airport functioned: "By plane, you step fully enclosed into one of the airport's 'airside' buildings and hence into one of its horizontal elevators to its 'landside' terminal, a 40-second climate-controlled ride that should be, and probably will be, a tourist attraction and a free way to spend a Sunday afternoon the airside buildings which were paid for by the airlines." B. O'Donnell (1991) described the airport in the *Tribune* as "laid out like a hub and spokes. The central building, or 'landside,' is linked by people movers to five satellites, known as 'airsides,' with room for one more airside. When further expansion is needed, a duplicate landside–airside complex will be built to the north." When the terminal opened for a public open house the weekend before flight operations began, 70,000–80,000 people clogged local traffic to tour the futuristic airport that was given a human touch with $1 million in landscaping. But with its efficient routing and passenger flow, the *Times* proudly predicted that "that was destined to be the last traffic jam to take place at Tampa International Airport" (Garret, 1976).

In another 1971 article, "Tampans Dazzled by Their Airport," Howard Gorham observes that:

> an estimated 20,000 to 25,000 Tampa Bay residents came and saw and apparently were conquered by the new Tampa International Airport yesterday. The authority threw open the doors yesterday to the modern new complex and let Tampa Bay residents get their first looks. The crowds came in a steady stream all through the day from 9a.m. to 6p.m. Officials had feared they might come in bunches and clog the building and its parking facilities. People wandered all over the complex, viewing first the main landside terminal, then venturing out on the shuttle cars into the four satellites. Many of them, it developed, frequently fly out of Tampa and were coming by to check out just what procedure to follow to get to their planes. The crowds came in a considerable mix—children, young married, elderly, middle aged. And all, obviously, were impressed.

However, the official fliers and pendants from HCAA recommended that users familiarize themselves with the new facility because "the more you learn now the less you'll have to learn … the hard way. Instead of risking 'gridlock' at curbside, park conveniently in the Short-Term Garage, and proceed to the attractive ambiance of America's most popular airport" (Tampa International Airport, 1972).

"The crowds came in a considerable mix—children, young married, the elderly, middle-aged. All obviously were impressed," was the way the *Tribune* shared its excitement. The local paper published comments picked from the public:

> The décor is beautiful. The color combinations are pleasing. It is restful to sit here. You can enjoy your wait.—Mrs. Earl Phillips

> We are taking a Cook's tour to see where to go. It's better than O'Hare (Airport) in Chicago. You've got to walk 10 miles to get anyplace there.—Fred Schmidt, of Tampa

> It's elegant. I was in Chicago (O'Hare) two weeks ago and this one has got it beat all the devil, as far as beauty and convenience. The décor is gorgeous, it is beautiful and comfortable.—Albert Straughn, 118 Cotton Ct., Auburndale

> I like the way you can park and go inside. The décor is wonderful. It doesn't look like an airport.—Mrs. Fred Schmidt

It is intriguing to learn how the first users of the airport are concerned with comfort rather than modernity. The first interviewee speaks of a restful and enjoyable wait; the second and third compare the short walks with O'Hare; and the third highlights that it does not look like an airport. Considering the positive response, it seems that the airport was in fact doing things right. In another article, "New Airport Jammed on Open House: They Came in Greater Thousands," Gorham (1971a) chronicles how thousands visited the new complex: "Airport officials said the giant terminal handled the vast crowds with great ease. And the viewers, as they were Saturday, seemed greatly impressed with the new terminal." Other commentators in the printed press seem to be very pessimistic with the current airport offerings around the country.

> Airport enthusiasts consider the prime function of an airport, in addition to the obvious, to be "to make a good impression." Most air travelers and airport visitors, however, would settle for less, a lot less. They would be happy with an airport that fails to make a bad impression. These are the airports that stand as the pride of other communities, an area of giant jams, invariable delays horrendous aggravations that local citizens diligently learn to avoid. Other airports are so terrible in appearance that they fade into being picturesque. But most airports have a non-look about them. Although all airports are places to pass through, these in question are left without leaving any sensation of having been anywhere at all … I advise to go to its "landside" terminal, a 40-second climate-controlled ride that should be, and probably will be, a tourist attraction and a free pleasant way to spend a Sunday afternoon. ("An impression," 1971)

I have to confess that I rode myself back and forth for many hours on that same People Mover, or at least that is how I remember what it was called during a 1978 school trip to Tampa. I can personally confirm that it became a tourist attraction.

Several opinions were sensitive to the reasoning behind the Landside–Airside Concept. The *Tampa Tribune* said that the airport:

> "... has a unique plan for eliminating "terminal sprawl"—an affliction for airports and passengers throughout the nation. What the term means is that passengers sometime have to walk thousands of feet—7,700 in one instance—through sprawling terminals to get from their cars to their planes, or vice versa. The problem is that in addition to physically wearing down travelers, the huge terminals were reducing the time saved by passengers flying the swift new jets. Lengthening the airports solved the problems for the aircraft but it left the weary passenger with more and more space to traverse getting to and from his plane." ("Terminal sprawl," 1971)

These radical changes were seen with delight by many and with an unleashed optimism by others. For a popular innovation, it reached such a scale that people thought they were witnesses to a new kind of airport, the airport of the future. Some referred to it as a "first generation airport" because of its odd looks and thought that, in essence, "Tampa Airport was dramatically different—futuristic, beautiful, efficient, state of the art" (Tampa International Airport, 1972). An enthusiast expressed after visiting Tampa that "airports symbolize adventure, pioneer spirit, freedom, exploration, change. In many ways, airports are a microcosm of the world" (Hill, 1996).

Tampa International's new terminal would become a posh, polished facility, "but it didn't get that way overnight … it took hard work and lots of imagination and money" ("Getting there's easy," 1971). Bean, the airport administrator, was prompted about the origins of Tampa: "Leigh Fisher, an airport consultant who advised officials on all aspects of planning and operating an airport back then, dreamed up the concept of having a main building where ticketing and baggage check- in occurred and a separate building … airports all over the world are now using (this) system" (as quoted in Brown, 1994). Other officials and politicians externalized their views of the airport. Tampa's Mayor Dick Greco (as quoted in "A Santa in April," 1971) described the building as "futuristic enough to make it a functional—and copied—for years to come." He pointed out that airports often are absolute the day they open because construction programs were not flexible enough to utilize new ideas and materials. And Hillsborough County Commission

Chairman Ellsworth Simmons believed that "the opening of this airport expresses complete confidence of this area in the development of the science in aviation." This observation is a classic illustration of the reliance on scientific progress in the American political rhetoric of the time.

As the accounts from travelers and journalists evidence, throughout the years TIA has won plenty of fans, including, surprisingly, the airlines, which in other cases usually complain about the smallest detail. An article from the local newspaper describes how "the airline industry itself represents another group of TIA customers that seems to have nothing but praise for the facility. Airline representatives applauded the design, operation and maintenance provided by the airport" (Taylor, 1986). This may be considered a breakthrough achievement considering the airlines' initial opposition to the project.

Last, the "Landside Airside" airport was far from romantic or aesthetically pleasing. But many accounts show that the grandeur of Tampa's airport would come from the socio-technical appreciation of its "revolutionary" concept. Some local newspapers showed the overexcitement of the moment. "Paris has a tower, London a palace and New York a certain skyscraper. Visitors are expected to see these landmarks. They haven't really seen San Francisco unless they've eyeballed the Golden Gate Bridge. Tampa now has its new air terminal. For grandeur, it can't compare with, say, the Coliseum in Rome; for beauty, it is no march for the Gateway Arch in St. Louis; for size, it isn't in the same league with Kennedy International" ("Tampa's new crown jewel," 1971). A double-page-spread advertisement by the Tampa Hillsborough County was published at the *Tampa Tribune* in 1971 (see Box 1).

Other companies, such as Westinghouse Electric Corporation, the main contractor of the ride system, thought that "after Tampa, airports will never be the same. No wonder airport people from all over the world are studying Tampa. Its designs will influence the 21st century. The design (see Fig. 2) was based on a landside/airside configuration allowing more effective use of aircraft" (Westinghouse Company, 1971). (Note that for the company the end beneficiaries are the airlines and not passengers, as officially stated.)

Fig. 2 People movers remain in operation after thirty-five years of its construction. Howes, James G. (2009). *People mover connecting Terminal A to the central terminal* [digital] (Reproduced with permission of James G. Howes)

Box 1 Advertisement for Hillsborough County Aviation Authority, Published in *Tampa Tribune*, 1971

> **The Future!**
>
> **Thanks ... to men of vision**
>
> **Thanks to the men of vision who plan and prepare for the unlimited opportunities of the future!**
>
> **Thanks to the builder ... who materializes those plans ... and builds for the future! ...**

A few months before the airport's opening weekend, the *Times* published an advertisement from Eastern Airlines (see Box 2).

Box 2 Advertisement for Eastern Airlines, Published in *Tampa Tribune*, 1971

> **"In October 1971, Eastern Airlines will fly to another world."**

The design life of the original system was twenty years. As stated in the "Tampa Airport System Replacement Program," experts explained that since 1972, "two additional airside buildings have been built with their own shuttle systems. In the 23 years since the initial installation, the vehicles and controls have served well, exceeding the design life of the systems" (Culver, 1996), although other companies tried to improve Tampa's People Mover, including Otis Elevator Co., who installed cars pulled by cable at Cincinnati's Airport and Tokyo's Narita Airport; rail-car giant Bombardier in Canada (who also builds monorails); and AEG Transportation Systems, a German manufacturer that talked about using experimental magnetic levitation technology someday for an airport train.

The *Tampa Tribune* noted, almost twenty years after the opening of the airport, that "the Westinghouse vehicle first tried in Tampa remains the American standard" (Taylor, 1986). Tampa's cars were "the first batch of automated people movers ever installed at an airport. Now, they will be the first retired. Officials at manufacturer AEG Transportation Systems think the early cars belong in a museum" (Huettel, 1995, p. 24). Museum pieces or not, the vehicles helped spark a revolution in airport design that saved billions of travelers untold trillions of steps.

Some twenty airports worldwide now have some type of people mover to whisk passengers between ticket counters, airline gates, baggage carousels, and parking garages.

Travelers at Chicago's O'Hare Airport could end up hiking two miles from end to end. The solution, concluded airport consultant Leigh Fisher, was to move the airplanes away from the rest of the airport. But there was a hitch. How do you move passengers quickly between the main terminal and remote airside buildings? (Huettel, 1995). Twenty-five years after its implementation, Tampa's journalists were still interested in the landside–airside system, and they continued to compare their emblematic airport with others, such as the disgraced O'Hare.

Tampa International Airport has become, over time, a distinctive part of the city. Even well into the twenty-first century, the "Landside Airside Concept" is still appreciated and in use. In 2011, CNN Travel published a list of the 10 of the world's most loved airports, and surprisingly, among hundreds of options, Tampa remained the sixth favorite, globally (Rane, 2011). Furthermore, in the national J. D. Power airport consumer satisfaction study of 2010, TIA was ranked second in the USA, among more than four hundred facilities.

However, not all withstands time; among the few features that are outdated are the congestion in the parking garage due to the lack of more ample drop-off and pick-up areas (which led to the "load and leave" policy in February 1995), the general aging of the prefabricated concrete buildings, among others; but surely the most sensitive point has been security.

The "Landside Airside Concept" (see Fig. 3) did not adapt well to the wave of security-driven changes after September 11. The physical, jurisdictional, and procedural constraints of the original layout have occasioned multiple pressures on the buildings and the designed passenger flows. In the words of Tampa's planners, there were basically no security zones at all in the former airport plan. The following three sections of this chapter are based solely on primary sources, organized around a number of interviews, plus the analysis of the original documents submitted by LF&A to the client during the 1960s. The LF&A team included Leigh Fisher, Marge Brink Coridan, Eugene Lewis, Jim Jones, and, later, Alistair Sherret and William Dunlay, among others. From HCAA, both Herb Gottfried and George Bean have passed away. Nonetheless, Marge Brink Coridan and Gene Lewis were the key planners of Tampa, and both kindly accepted to give long interviews for this work. Both participating in later stages at TIA, Alistair Sherret, a business and financial airport planner, and Bill Dunlay, an expert on airfield and airspace planning, also generously spoke at length about these topics.

3 Problem Posing

In the case of Brink Coridan and Lewis, much of their expertise in airport planning came from the practice. By the time they both became involved in TIA in the early 1960s, the former planning team, with

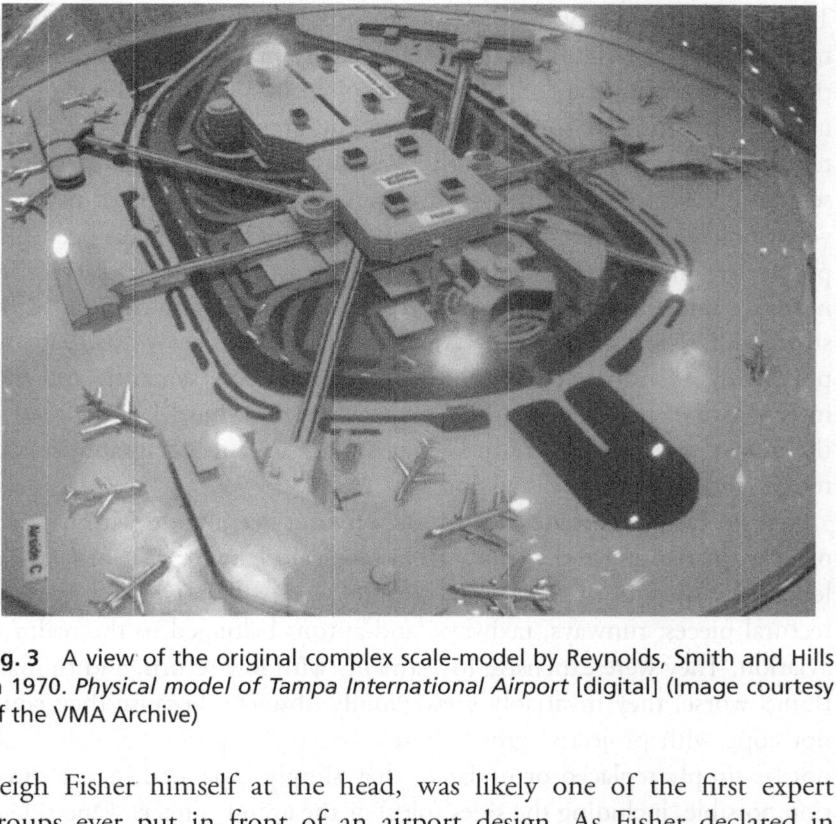

Fig. 3 A view of the original complex scale-model by Reynolds, Smith and Hills in 1970. *Physical model of Tampa International Airport* [digital] (Image courtesy of the VMA Archive)

Leigh Fisher himself at the head, was likely one of the first expert groups ever put in front of an airport design. As Fisher declared in 1965, aviation authorities were unable to figure out a solution for airport congestion. In fact, what FAA asked to the planners was to pose the problem itself.

> It's the first time in all my 18 years in the consulting business that this has ever happened. Everybody wants a terminal right now, maybe even yesterday. They define the problem and say "Find the solution." Tampa has in effect reversed that procedure and told me to define the problem first.
>
> —Leigh Fisher (1965)[3]

[3]Hillsborough County Aviation Authority. (1965). *A new approach to jet age air terminal development: Progress Report 1961–65*. Tampa International Airport.

This meant that a new method of analysis was necessary. An architectural approach to the problem would usually materialize a building; in this case, however, the collaboration resulted in a plan, or the diagrammation, of future actions that inform the architectural program of a terminal. In planning, the creative work is diagrammatic and strongly semantic. The following paragraphs present full fragments of my conversations with members of this group of planners; they have been purposely left untouched as whole paragraphs in order to preserve their narrative and idiomatic force. Therefore, my analysis spins around the strong differences among experts with respect to the keywords that support their individual views and ideas. It is decidedly semantic and thus interpretive of the specialized language expression. Last, I underlined in the quoted paragraphs certain keywords, jargon, and paragraphs related to the landside–airside discussion.

The problem was evident: Airports were not keeping up with the pace of growth, passenger demands, technological changes, and other challenges. By the end of the 1950s, airports were still thought of as architectural pieces; runways, taxiways, and aprons belonged to the realm of aviation. They were expensive investments with low return, and to make things worse, they invariably grew rapidly obsolete, because they could not cope with projected growth rates. Because airport terminals could not be simply replaced or updated, they simply expanded in any direction possible, including the sides (pier) or the apron (finger). One side of the airport was subject to the pressures of technological change (aircraft changing dimensions, airspace slot patterns, noise, new codes, etc.); the other was simply not delivering what people wanted. When LF&A was called to devise a concept for Tampa, according to the HCAA (1965), the team decided to undertake the "research path and Fisher was given time to explore a variety of possible solutions to the terminal dilemma." But it was not until LF&A's Jones, Brink Coridan, and Lewis identified and verbalized these issues in planning terms that the planning problem was finally posed and discussed. LF&A internally agreed on defining five keywords, which literally launched the final planning report: landside, airside, terminal, conventional Terminal, and transfer device.

(Note that the words landside and airside—with the exception of their profuse mention in ICAO's *Annex 14*—were no longer used at the time.)

LF&A's final product, the *Recommended Planning Criteria for New Passenger Terminal Facilities at Tampa International Airport*, begins with a glossary (see Box 3).

Box 3 Glossary from ***Recommended Planning Criteria for New Passenger Terminal Facilities at Tampa International Airport***

As set forth in Volume I of this study, the design of any airport terminal area must reconcile the requirements of three basic operational areas:

Airside—	The area where **aircraft** loading and **servicing** take place, including the passenger processing space dictated by any given design.
Passenger Collection Point—	The facility or facilities where passenger processing and service take place prior or subsequent to passenger transfer between airside and landside (conventionally it is the terminal building).
Landside—	The area where all **ground** **transportation** requirements (roadway systems and parking areas) are accommodated.

Airside functions

The physical layout of the aircraft parking and servicing apron for an airport. Must relate passenger handling procedures to aircraft servicing requirements. Generally, co-mingling of passengers with ground servicing equipment and aircraft should be held to an absolute minimum. The physical configuration for any given airside design is dependent upon both the number of gate (loading/unloading) positions required by the airlines and by each individual airline's methods of servicing and handling its aircraft at the gate position.

Landside Functions

Surface transportation is the major consideration and problem area on the landside. The provision for rights-of-way, properly located curb frontages for off-loading and pick-up, and holding and parking facilities are required for all modes of ground transportation ranging from private cars, taxis, limousines, busses, and rent-a-cars to trucks delivering cargo, mail and express as well as those servicing the airport with goods and commodities or removing trash and debris. All systems providing ground (or helicraft) transportation to airports require roadways or surface rights-of-way (ROW) on the landside of the function. These vehicles may be broken down into two categories:

CONVENTIONAL **TERMINAL** CONCEPT

In a conventional terminal concept, airside requirements have almost always been contiguous to the passenger collection point (terminal building) and connected to the passenger processing facilities by some type of finger or concourse. In a conventional concept, the landside requirements have likewise been contiguous to the terminal building and have usually been situated on the side opposite the airside.

The passenger collection point (terminal building) has historically stood between the landside and airside both as a funnel to get maximum concession exposure and also as the single link which houses the transfer of passengers and cargo from surface to air transportation. This degree of centralization to facilitate landside convenience, coupled with the extensive decentralization demanded in the airside, has caused excessive walking distances and the loss of proper human environmental scale in most conventional terminal solutions.

Note: Reproduced from *Recommended Planning Criteria for New Passenger Terminal Facilities at Tampa International Airport*, vol. 2, by Leigh Fisher and Associates, Inc., July 1963b, San Francisco, CA: Author. Emphasis added. Here Brink Coridan describes her views of the landside and airside.

The above definitions are certainly consistent with Marge Brink Coridan, who authored the document. However, they do not necessarily coincide with some of her colleague's definitions of the same terms. According to Brink Coridan, landside and airside establish a machine–machine relationship, which captures the essence of the plan. For her,

> The landside infrastructure held the permanent building, while the airside has only temporary buildings, supposed to be torn down along with the airplanes changes, and mostly for money reasons. (Personal interview, M. Brink Coridan, March 8, 2008)

In short, for Brink Coridan the airside is the zone of aircraft, and the landside the zone of the automobile (or other terrestrial means of transport). In between, there is a passenger collection point (PCP), or terminal. The archetype of the terminal or station too often makes reference to monumentally epic spaces, where the steam of trains let us know they are ready for departure. The romanticism of this imagery strongly contrasted with the severe austerity of how the "Landside Airside Concept"

Fig. 4 View showing the mechanized connector between the landside and airside buildings. *Bird's-Eye View of Tampa International Airport* [digital] (1973). Public Domain (Provided for reproduction by the Florida Photographic Collection at the Florida State Library)

was finally materialized. The glamor of transportation of terminals was now reduced to dark corridors, escalators, and small carpeted rooms. Thus, the word *terminal* needed to be removed, at least as a reference. The way keywords are organized in the above text makes the real nature of the enterprise more transparent. What LF&A was trying to achieve at this point was to materialize a diagram: a diagram understood as a chart, which describes the logical sequence of a system or routine. What Brink Coridan was looking for here was to pose a problem through semantics. Should she use the term *terminal*, the complications of negotiating between human and machines would be far higher, so a new term must be coined instead. The "Passenger Collection Point" is a term or a label that perhaps deflects possible criticism for having shrunk the terminal to a miniature version—removing all possible *grandeur* and thus becoming anti-*cathedralesque*. The *semantic diagram* deconstructs the terminal in several pieces. Users work as a bridge between two poles: two technologies that ask for different requirements.

The diagram poses one serious problem, though, as stretching the boundary would imply increasing rather than reducing the distance between passengers and aircraft. The new boundary was initially called the "ride system," later the "shuttle," and, finally, the "transfer device" system—which needed to be made to fit. The system is defined in the document as a "tool to serve a function." Incidentally, this artifact was known humorously among the design team as the "horizontal elevator" (see Fig. 4). Years later, and after the collaboration with Westinghouse, it was finally baptized the "People Mover."

Returning to the landside and airside definitions, as Dunlay rightly expressed it in an interview: "You have to be very careful every time those terms are used or people would interpret them differently." When Alistair Sherret, who is a Cornell engineer and who served as a financial/operational planner of LF&A, was asked about the Landside–Airside Concept, he spoke of balancing capacities:

> Airside being the airspace, airfield, taxiway system, and the landside being the gates and the terminal and ground transportation. So the basic question to be resolved is the balance of the capacity of the airside versus the landside. And ideally those two would be in perfect balance and there would be sufficient capacity and there wouldn't be delays. But in practice there is always an imbalance. (Personal interview, A. Sherret, March 8, 2008)

Sherret does not see the terms as zone-defining, but rather as transportation subsystems, almost as interdependent subcultures. Among them, he speaks of *delay* as the foremost enemy. His use of the word *versus* apparently speaks of a silent antagonism. Later, he mentions *balance* as a solution to this conflict: A perfect balance could make delay disappear. Sherret is tracing a line, a path. Dunlay, a distinguished Berkley and University of Pennsylvania professor, rapidly responds when he is questioned about the terminology:

> One thing that I would note that ... These terms are not used as correct or as precise. It may suggest a boundary that is different in different airports and different people ... The original idea at Tampa was mostly the airside was the concourse where you park aircraft, and the landside was everything else, and the terminals, specially ... But many people would make the distinction between airside terminal and landside, strictly the access roadway and parking so ... I take it here the landside is where the terminal ends and then the access parking, and the airside is where the aircraft park and passengers board ... The definition is a little fuzzy ... So each time you use those terms you have to define them ... Usually people refer to landside, people at FAA and people doing research and the National Transportation System ... The general thing is that airside has been from the apron out. And then terminal and everything with passengers and ground vehicles, that's landside. In the classic Tampa split, the definition was to separate one part of the terminal from another part of the terminal: the airside part of the terminal where the passengers board the aircraft, and the landside portion with the ticketing and the baggage claim, typically that included the curbside and the access roads too, split in two. You have to be very careful every time that term is used, people would interpret it differently. (Personal interview, W. J. Dunlay, March 8, 2008)

For Dunlay, these terms are clearly "fuzzy." For the expert eye or not, his adjectives may be not that different. When the airport engineer speaks of "different" boundaries, he makes me think of *changing boundaries* or frontiers that are in constant movement, either historically, physically, or contextually speaking. That would be congruent to the need to define them on a *per-case* basis and would be explained as interpretive flexibility.[4] It makes sense to think of different relevant social groups providing different interpretations of the "Landside Airside Concept" when, at bottom, they all picture different solutions for the same problem. What is even more striking is that Dunlay believes that people—clearly referring to planners but perhaps addressing users—would interpret it differently. How can you base an "airport revolution," as described by the HCAA (1965, pp. 1–2), on such ambiguity?

[4]As in SCOT (Social Construction of Technology; Pinch & Bijker, 1987).

Marge Brink Coridan is the first female airport planner on record, and she has been considered for decades an authority on the matter. However, her vision of landside and airside is far from standard or encyclopedic. After reading multiple documents regarding these projects which span almost a decade, I have found that she was consistently congruent with her initial definitions and that those definitions remain almost intact nowadays, even after a lifetime of practice in the planning profession. When asked about the linguistics of airport planning, and more specifically landside and airside, Marge Brink Coridan responds with a nostalgic expression on her face:

> My definition of airside is everything that has to do with airplanes, while landside is everything that has to do with ground transportation. In this definition the Terminal building is just a connector, not being part of either the airside nor of the landside, but primarily containing the "ride system." (Personal interview, M. Brink Coridan, March 8, 2008)

For the legendary planner, the landside and airside are defined by a conflict between machines only, rather than a relationship between human and machines. During the conversation, Brink Coridan emphasizes that the terms were coined with only transportation systems in mind, around the spring of 1961. She argues that the historical conflict within airports has to do with the unresolved problem of making efficient connections between two different systems of transportation: the airplane and the car (or other mass transit systems). Regarding the specific use of those terms in Tampa, she holds that "in this sense, the landside infrastructure held the permanent building, while the airside has only temporary buildings" (Personal interview, March 8, 2008). Here, Brink Coridan reveals another key idea. She implicitly defines airside as temporary and landside as permanent. This idea may have several interpretations, even in philosophical terms. The reasoning behind it according to her comes from the very nature of the way we have constructed and evolved our large transportation systems. She holds that terrestrial transport, such as the automobile, has suffered very modest changes since its origins; thus, there is basically no need to change or adjust the infrastructure, which facilitates the flow or parking of vehicles. On the other hand, there is constant change in aircraft (particularly during the postwar period), thus its transformative nature.

Brink Coridan (Personal interview, March 8, 2008) last considered in the interview "the relationship between landside and airside to be practically the same, since they are not actually separated," and for her, the boundary is not a boundary but a *case,* since aircraft and passenger services are both servicing functions. If she had to draw a line, it may be across the face of the terminal building, which is perhaps closer to what many planners in the business would think. In the end, I would consider Brink Coridan an unbiased mind, able to associate things that others perhaps could not, and audacious enough to label them with "new" words.

From the opposite perspective, a known rebel without a cause and long-time Bay area inhabitant, Gene Lewis, is still an active planner and a legend in the business. I have had the pleasure to work alongside him and to learn much from his incontestable planning expertise, but beyond that I am privileged to enjoy his friendship. He is a great conversationalist and a true enthusiast of airports. Our talk is loud, and Lewis passionately responds to the key question of this section: For the former leading planner of Tampa, and now several decades after, what is the meaning of landside and airside? Lewis has to respond twice to the same question, as our interview must be split into two parts. During the first part of the encounter, Gene speaks freely and in a more improvised manner, and, gesticulating, he begins to make rapid connections:

> Airside? But it's still not a real word. It's more important what's inside. Every time you type it up, there is no such word as the airside. It isn't really a popular medium. Since the airports are so small, to the general population, nobody else uses the word except the airport planners, and the airport people. And that's almost nobody ... So, when do you think that started? The use of the words? I think it started by ... it started definitely with Tampa, people started using it, just getting to the idiom. Remember it's only used by airport people. If you take everybody working at all the airports, and the airline personnel and sort of airport planners, and engineers, and architects, and stuff like that, that's a very small population. Because how many towns have more than one

airport? Almost none. How many towns have no airports? There are only 200 airports in the United States. How many hospitals are there? How many schools? How many churches? There is hundreds, thousands, hundreds of thousands of those. There are only 200 airports. And each airport has a relatively small number of people that keep it running 24/7. That's not a lot of people. And most of the airports anywhere, they don't call it landside and airside. At Tampa, the name was supplied for two different side buildings, it became part of the graphic system. Airside A, Airside B, Airside C and Airside D. And then while they were given the name, there had been airports where they had airside and landside, like Atlanta is a good example, the busiest airport in the world! Most passengers were landing there than any airport in the world. They didn't use the term landside–airside. They used the term "concourses." Concourse A, Concourse B. As soon as you start giving this stuff these weird names that no one knows about … Concourses came from trains. Train stations have concourses that are what they are concourses. Trains would come in on a track, like this (moving both hands in parallel) and each one of these tracks had a walkway in between and that was like the concourse. I don't really know why the term concourse, it sounds kind of French to me. But most airports call it Concourse A, Concourse B, Concourse C. This is a term only airport planners use. Other than read the signs where it says Airside A, Airside B. In the heat, the people mover was given different names: the ride system … they keep giving it different names: the shuttle this … the transporter this … and all that stuff is confusion. You have to come up with some kind of term, at least for airports, because everybody has to know. (Personal interview, E. Lewis, March 7, 2008a)

Lewis affirms that the terms do not exist outside the limits of the airport world, and that therefore they represent a sort of "expert code" in terms of meaning. However, he also suggests that these terms give Tampa its own sense of identity, and that this identity was embraced, but not necessarily understood, by airport users. This idea is reinforced by the fact that, in order to avoid becoming a copy or a derivative, Atlanta sought its own language, its own idioms, interchanging Tampa's "landside" for "concourse" (see Fig. 5). Did airside–landside become Tampa Airport's trademark for posterity? According to Lewis, very few airports call it landside or airside anymore, but why? Last, Lewis leaves clear that the terms landside and airside define buildings as orientation

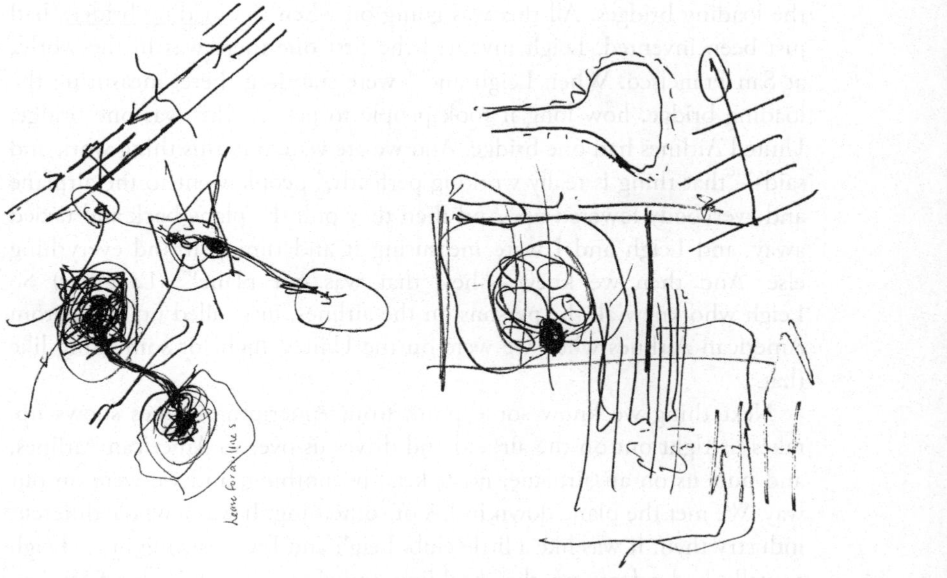

Fig. 5 Gene Lewis' sketch illustrating how Tampa's concept of radial satellites—shown on the left, evolved into Atlanta's idea of a sequence of modular terminals. The author makes emphasis on the mechanized connections in both cases. Eugene Lewis (2008). "From Tampa to Atlanta" Ink drawing on paper (Courtesy of VMA Archives)

points or references within the airport geographic system. From now on, he refers to these boundaries as physical barriers.

During the second part of the interview, which took place just a few hours later, Lewis appears much more focused and reflective—by his facial expression it is easy to guess that he is making a strong effort to recall events as far back in time as he can:

> The landside–airside boundary was literally where the airfield started. Everything else was available to the general public. Before they had real security like they've got now, the landside–airside was always secure. You couldn't get out on the airfield, anywhere near the airplanes. You couldn't get on the airplane without going through ... You weren't able to walk out to the airplanes ... Somebody from the airlines, when it would be reported, now it's time to get on the plane, they opened up the door to

> the loading bridges. All this was going on when the loading bridges had just been invented. Leigh invented the first one there was in the world, at San Francisco. When Leigh and I were standing there, measuring the loading bridge, how long it took people to get … This was one bridge, United Airlines had one bridge. And we are watching this thing work and said—"that thing is really working perfectly," people went to the airplane and everybody lowered up. And then they pull the plane back and taxied away, and Leigh and I were measuring it and timing it and everything else. And then we knew—"hey, that was our plane!" (Laughter) So Leigh who knew all the persons on the airlines, just called up a guy from American Airlines when we were on the United flight or something like that.
>
> Next thing we know some punk from American Airlines shows up, takes us right out on the airfield and drives us over to American Airlines, and puts us on an airplane, no tickets, no nothing, and we were on our way. We met the plane down in LA or something. It was a whole different industry then. It was like a little club. Leigh and I were sitting in … Leigh actually had a lease on that building and they were there for 55 years. He basically owned the airport for 55 years. (Personal interview, E. Lewis, March 9, 2008b)

In this second part, Gene continues to express his previous ideas, but in addition to supporting historically why he thinks the boundary is mainly physical (he even refers to it as a "door"), he introduces a new concept: security. In this deeper meditation, Gene depicts the landside–airside boundary as a way to secure two territories. This border is material, and passengers need to *go through it*.

I find particularly interesting the fact that he uses that expression, as if the boundary would have enough depth in order to go *through* and not just *in* or *inside*. Last he associates the air bridge with the boundary. This time he imagines the frontier as a passage, which connects two edges: the airplane and the terminal. But in the end, he recalls fondly "it was like a little club."

In this section, I have argued that the main planning problem was isolated and posed, once the terms landside and airside were dusted off and put into a new context. The terms' meanings were not agreed upon among the key actors as one may suspect; rather, there was a consensus on the *semantic diagram* or, in other words, the way the terms interrelated with each other, forming a sequence. This *semantic diagram* (see Fig. 5 and other relevant documents and videos in www.landsideairside.com) is an argumentative solution to a historical controversy: the way airports in the 1950s were laid out, as a model, does not *progress* at the same speed aircraft does (technologically and economically). And that is precisely why the identification of the landside–airside boundary is one necessary step in the understanding of why the airport—seen as a technological object—is socially constructed. The term terminal had to be replaced with the newly coined PCP. The outcome of such *idiomatic debate* may be illustrated with the LF&A diagram shown in Fig. 6.

Again, in my view, the fact that at the time of posing the problem experts could not agree on a word's meaning sounds natural and could be explained as interpretive flexibility. As the artifact has not yet materialized, it only holds everyone's opinions and interpretations in linguistic terms. It is almost an abstract idea, but a powerful one. *Let's split the airport, and then, let's reconnect it again with a machine!*—they may have thought. In the case of the "Landside Airside Concept," in SCOT terms,[5] the relevant social groups reached a "rhetorical closure" and agreed on the theoretical disappearance of airport planning problems.

4 Problem Construct

By the spring of 1962, airport planners had won the first battle. They had identified and labeled what most architects, engineers, and politicians had failed to address in the past: the rising conflict between the landside and the airside. Now, the HCAA (1965) faced a difficult decision between commissioning a "conventional" terminal and taking the more thorny "research path" (p. 9). Following the steps of FAA's initiative for Dulles International Airport, Herb Godfrey, HCAA's executive director, agreed with Leigh Fisher that a certain methodology was needed to study the alternatives and support the possible solutions.

[5]As in social construction of technology (Pinch & Bijker, 1987).

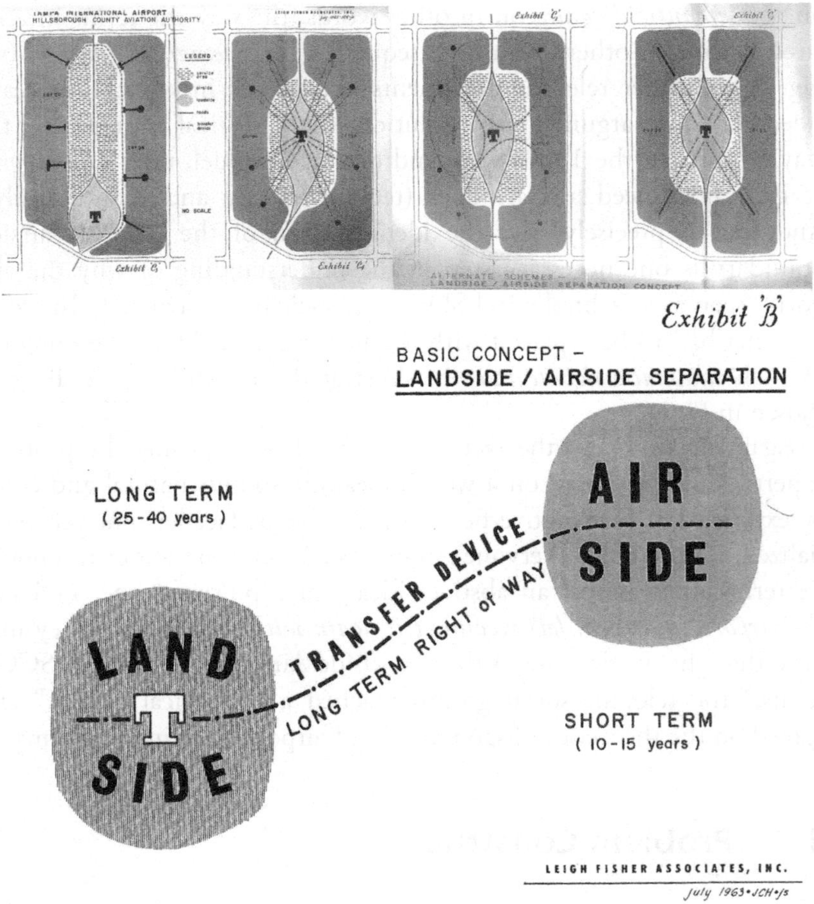

Fig. 6 These diagrams show the different possible layout arrangements for a landside–airside separation. Leigh Fisher and Associates (1963). "Basic Concept Diagram, Landside–Airside Concept" by Brink Coridian, M. (June, 2006) in Leigh Fisher Associates: History and Lore, as see through the eyes of Marjorie Brink Coridian (Courtesy of Marge Brink Coridian)

Thus, the original contract signed by all parties asked for a thorough "analysis of major airports in the United States today and a definition of the basic problems created by current terminal operational concepts" (HCAA, 1965, p. 7).

Nonetheless, the operational problems were visible to everybody in the industry, although they were not identified using specific terms that could make them understandable until Brink Coridan labeled them and those labels diffused. The first part of the contract, though, demanded a large-scale comparative study, just like the one Saarinen conducted to justify his concept at Dulles.

I argue in this section that the methodology's main objective was the construction of credibility rather than the provision of valuable information that could inform either planners or designers of new potential solutions. In spite of the tremendous traffic growth pressures on HCAA's airport management and the original target date for occupancy of 1966, the team had to face rising planning costs and a two-year minimum delay for research purposes. I deduced that LF&A was very confident, hiding a tentative solution in their pocket, before pushing their client to undertake a path where "the dangers of miscalculation and unfavorable repercussions from the community if the transition from time to the present terminal to a new facility was unduly extended" (HCAA, 1965, pp. 5–6). In this regard, Gene Lewis recalls the following:

> The technical committee from the airlines that was involved, we had the airport planning consultant, which is Leigh Fisher Associates, the majority game was Leigh and Marge Brink,' cause she did all the studies. And then you have the airport director, two successful airport directors. The key one was George Bean, who was the airport director during the final stages, and most of the stages of the planning. … Godfrey was involved initially. And the engineering firm Grinder, which is now URS, they had already designed the improvements to the airfield and kind of a general layout for the apron area, where the aircraft would park, kind of. … And there was a scheme of conventional pier plan , you know, finger plan . … which had been designed based upon on what had already been done, for airports of the size of Tampa, with few exceptions. … so when someone along the line. … they decided, Leigh and the director and everybody, except the airlines, it was suggested to the airlines, we won't explore these other concepts. … They decided to make Tampa the most convenient, passenger-friendly and efficient airport they could possibly think of. … and that's when they got into. … when Marge did that survey, the

> pictures I've showed you on the wall. … Marge did a survey on all the airports, in the United States, not all of them but the most significant ones in terms of the broadest possible range of different kinds of concepts. … you know, the flights at the gate. … etc. (Personal interview, E. Lewis, March 9, 2008b)

The "other concepts" Lewis mentions were in fact the landside–airside early hypothesis. In agreeing on a method, the participants in this project gained something that, building on Bijker (1997), could be called problem framing or a problem construct. In my view, this phase follows inevitably after posing a problem, or in other words, after launching a design hypothesis, which is necessarily semantic. In framing, a problem will be simply put in context. As there is no such a thing as "scientific truth" in design or planning matters, certainty must be reached by consensus. Thus a design hypothesis could be framed by comparisons and analysis of one or more of the following:

- Current styles (conventional designs, new design trends, etc.)
- Historical performance (measured levels of efficiency in previous, similar cases)
- Risk (comparative success/failure rate among historical innovations)

In the case of Tampa, these three aspects are fully covered in the document *Study of U.S. Airport Terminal Buildings as a Basis for Planning New Terminal Facilities at Tampa International Airport*, a report by LF&A in April 1963, that consists of two volumes,[6] including a table of contents, which divides the studies into subsections. In my view these subsections address the three aforementioned categories in the following manner:

- Current styles are analyzed in the "Evolution of terminal building design."

[6]It is in fact somehow confusing that the two report-books have been labeled volume I & II, while they differ in title and contents and their publications are separated by three months. On the other hand, volume I contains the "basis for planning" and volume II the "planning criteria" itself, forming in a way a single integrated study all together.

- Historical performance is addressed by the larger analysis of selected US airport terminal buildings and a special section devoted only to provide a critique of Dulles International Airport.
- Risk is analyzed in the evaluation of current types of operational plans, which also includes comparisons between investment and revenue among different airports.

LF&A's "framing of the problem" concludes by showing "evidence" that definitely supports their initial hypothesis: The landside and the airside must be split. The document, which is over two hundred pages long, reports partial findings per section, being either diagnostics, comparative results, critiques, or evaluations. A thorough analysis of the report provides many traces of the LF&A team's process of thinking after interpreting the data. In the following section, I discuss and summarize the team's findings associated with the construction of the "Landside Airside Concept."

For LF&A the nature of the airside is variable, and the nature of the landside is constant. A paragraph in the report addresses this issue:

> … the length of the curbside of the terminal building has remained relatively constant over the years, irrespective of increasing passenger volumes and aircraft activity. As a matter of fact, the entire "land side" structure has not changed to a great degree over the years, regardless of the amount of passenger traffic. The reasons for this evolution in terminal building design are clear.
>
> Twenty years ago, terminal sprawl did not exist simply because air traffic was nominal. Ten years ago, with the air traffic volumes rising sharply, the need for expanded facilities was beginning to be recognized, but airport-financing ability was generally lagging behind the increasing demand for air transportation. (LF&A, April 1963a, p. 3).

This idea unfolds the strains between the two sides of the airport. It speaks of different speed of change, programmatic requirements, technologies used, and congestion. According to LF&A, the landside has remained immutable, and the airside is under pressure to cope with increasing demand. According to my interview with Brink Coridan, terminals have "evolved" in a way that is contrary to passengers' convenience. The comfort of passengers has been historically neglected. A report titled *A Study of Transportation Means Between Airports and the Metropolitan Areas They Serve* (Nordle, 1961), prepared by Human Sciences Research, Inc., for the FAA in 1961 states that surveys of the travel market characteristics in certain areas of the USA have indicated that "comfort" has been given as second only to "less time" as the reason for choice of transportation mode. Comfort, as defined by Human Sciences Research, is the degree of physical and mental ease experienced by a traveler. Elements relating to mental ease may be considered to include wait times, speed and ease of check-in, ease of locating flight departure room, and any factor that contributes to the comfort or discomfort of the passenger. G. W. Stallibrass, England's Director of Aerodromes, in a 1959 article titled "Trends in Airport Planning," said, "A recent IATA document rightly points out that internal design of the terminal building 'should reflect a concern for the convenience of the passenger.' Maybe convenience should not stop at physical progress, but at ease of mind as well" (quoted in Nordle, 1961, p. 4). Airport terminals in the late 1950s suffered from several problems that were analyzed by an FAA planning group before the Dulles planning phases. LF&A's group of experts "evaluated existing passenger handling systems," as follows:

1. The conventional (ground loading) systems make the passenger walk rather long distances in the fingers and additional distances to the plane, exposed wind, rain, aircraft noise, blast, etc. It often requires two changes in levels. It requires long and intricate aircraft taxiing, but does not demand precision positioning of the aircraft at the gates.
2. The loading bridge system involves a long walking distances in the finger, although all on one level, and it keeps the passenger sheltered at all times. It requires long and intricate taxiing and demands precision positioning at the gate.
3. The bus system, in use today in London, Frankfurt, Amsterdam, etc., cuts down passenger walking distances. It requires changes in levels and although the passenger is sheltered getting on to the bus, he is exposed to weather, noise, etc., in the transfer from bus to plane. The high steps, narrow aisles, inadequate hand-baggage space, etc., of conventional busses add to passengers discomfort. The bus system allows the aircraft

minimum taxiing distances and demands no precision positioning at the gate. While the bus system (European), as we know it today, does have disadvantages and problems, the very fact that it solves the walking distance problem, and at the same time minimizes taxi distances, makes it the most promising of the passenger handling systems and one that merits extensive study. (LF&A, April 1963a, pp. 34–35).

As I emphasized earlier, it is clear that the justifications for not following conventional airport schemes are fully based on improving passenger comfort (other reasons regarding aircraft are secondary). A maximum walking distance should be determined and agreed on. The LF&A (1963) study notes that, according to the "Port of New York Authority, a maximum walking distance should not exceed 1,250 feet," and although this has been the reference for years, of the airports studied by LF&A, Dulles International Airport has only 160 feet of walking distance, which is the lowest minimum walking distance for an originating passenger to the nearest gate from any airport in the USA. In contrast, the highest maximum walking distance to the farthest gate is 1,735 feet (O'Hare International Airport) (LF&A, April 1963a, p. 9).

As stated by LF&A (1963), revenue is a rising aspect of modern airport planning. The first volume of the study concludes that "the modern terminal building should be designed to maximize revenues from concessions. Hence, the emphasis on the 'forced feed' principle of terminal design—whereby passengers are 'forced' through concession areas on their way to gate positions" (p. 6). The report shifts its argument and critiques Dulles Airport. Certainly, Washington's international airport is a paradigmatic case, but several issues were overlooked in its commercial planning. For Brink Coridan the capital's airport "was designed specifically to reduce walking distances—and has done so—to a usual maximum of no more than 200 feet" (as quoted in LF&A, April 1963a, p. 13). The airport is certainly a "significant step in recognizing the problems of terminal planning and, as such, provides a necessary catalyst to provoke original thinking in terminal building design" (p. 39). However, although the Mobile Lounge has reduced passenger walking distances in the terminal building in most instances, it has also introduced certain new problems, as described in the report (p. 39):

1. An additional vehicle has been added to airport ramps. Apron maneuvering is delicate because of the size of the lounge.
2. Trickle loading of an aircraft is impossible. Once aboard the Mobile Lounge, the passenger must wait from 15-30 minutes before departure to the aircraft. Passengers are in effect held in "limbo" in the Mobile Lounge until they get to the aircraft—once aboard the lounge, they cannot take advantage of concessions.
3. Congestion at the boarding gate is increased. Passengers who have been accompanied by friends tend to cluster at the doors for last minute conversations.
4. The moving of passengers as a group results in "bunching" of passengers, thereby creating unusual peaks at escalators, circulations points, etc.
5. Passengers tend to queue up in the lounge as it approaches the aircraft.
6. Loading steps are still required for crew, service and connecting passengers.
7. The handling of connecting passengers is inconvenient. (LF&A, April 1963a, p. 38)

In conclusion, the problem framing furnished HCAA with convincing elements in order to undertake and finance the mammoth task of reinventing the airport. As stated in the progress report, the new concept is not a melding of the features of other major air terminals, but a "*totally new concept* of moving passengers from their ground transportation to their airplanes" (HCAA, 1965, p. 2). Based on LF&A's conclusive study, Director Godfrey asked the Authority's members "to abandon the avenue of normal procedure and strike off into unexplored territory—where amid a tangle of risks the real answer might be found" (HCAA, 1965, p. 3).[7] Officially, just two years after the inauguration of Dulles, the airport needed to be reinvented, and thus the problem was finally constructed.

---

[7]The cited *Progress Report: A New Approach to Jet Age Air Terminal Development, 1961–1965,* by the HCAA, was based on LF&A's previous studies (Volumes I & II). This document served the HCAA basically as a way to persuade the Board of the Aviation Authority, who would have the last word on the new scheme.

5 Problem Solving

Although an airport terminal may be somewhat analogous to machines or artifacts, it would be more accurate to say that the airport as a whole is a technical system within a technical system. Surprisingly, the history of airports shows that when *designers* are in control of the invention-innovation, the system builder approach by Hughes (1983) is an appropriate method of analysis. But when *planners* are in control of the invention-innovation, Bijker's (1997) technological frame fits better as a methodology for understanding both the "creative process of invention" and technological change (p. 123). This makes more sense if we think of these processes as strictly different in nature: Designers must often impose designs (objects) that will determine how users behave; on the contrary, planners will impose plans (not objects), and suggest in a non-material way how problems may be posed (verbalized), framed (contextualized), and later negotiated (agreed upon).

Tampa's gestation process is a good example of the above. Even when planners may be right in terms of identifying problems and solutions, they will invariably have to face the ultimate frontier: the negotiation. In these terrains, planners must be equipped with solid arguments if they are to resist the critical path awaiting them. Two central questions arise: By whom was the landside–airside boundary negotiated in Tampa, and how?

The members of the relevant social groups that participated in these negotiations "involved all the major stakeholders," according to Dunlay (Personal interview, March 8, 2008). For the HCAA (1965), all actors were "brought into the picture at the outset of the Terminal Design team's meetings" (pp. 1–2). The various governmental agencies involved include the FAA, the mass communication media, users of the airport, particularly the airlines, and the members of the design team: HCAA, headed first by Director Godfrey, and later by his successor, Georges Bean; the engineers J. E. Greiner Company, from Baltimore; the architects H. Leslie Walker & Associates, from Tampa; Thomas Sullivan, First Deputy Director of Aviation with New York Port Authority—who acted as an architectural technical advisor; J. Van Ingen & Company Inc., of New York and Miami, who were fiscal advisors; and the airport planners and main technical advisors Leigh Fisher and Associates, of San Francisco.

Unlike its predecessors, the boundary was encapsulated in one monolithic idea: Brink Coridan's "Landside Airside Concept." Consequently, and for the first time, the negotiation process was apparently reduced to a simple acceptance or rejection of the concept. Brink Coridan responds categorically to this effect:

> It wasn't negotiated. The concept was an airside–landside separation to reduce walking and improve passenger convenience; thus nobody won. The consequences were that the airlines were scared and the authority members thought it was wonderful. But the airlines had good relations with Leigh Fisher, so it became the basis of a prototype. (Personal interview, M. Brink Coridan, March 8, 2008)

Brink emphasizes that "Leigh Fisher could sell anything" and that the engineer was a real visionary, a man who tested every idea and had a lot of integrity. "When he presented the concept to the authorities and the airlines, they simply validated it," she ends. Although Brink Coridan is right in assuring that perhaps the planning concept did not find much opposition, the records show that this may have been one of the longest negotiation processes ever in the history of airport design.

For Alistair Sherret, negotiating landside–airside boundaries comes down to congestion and delay. During the interview, Sherret responds as follows when asked about the difficulties found in the process: "The tension in the planning process is the bias for the airport to emphasize service to the passengers from the bias on the part of the airlines to minimize their cost and maximize their operating efficiencies." This speaks of opposing interests from opposing stances: the airport at the landside and the airlines at the airside. Sherret believes that ideally those two would be in perfect balance; thus ensuring enough capacity and therefore no delays. "But in practice there is always an imbalance." He remarks, "From the perspective of airside: The passengers in the airplanes are the top priority and responsibility of the airport. To put it oddly, the airlines cares a lot about the airside objective: minimizing delays. However, they are not as concerned about delays on the landside unless it affects their airside operations. This lack of common goals creates an everlasting tension between landside and airside (Personal interview, A. Sherret, March 8, 2008).

According to Gene Lewis, the negotiations at Tampa were held in a context very different from that of contemporary airport design. Commercial airports were in their early stages, and less than 3% of the world population had flown on a plane. At the time, airlines were very powerful, but airports were, too, because they were strongly supported with taxpayers funds. Gene recalls that LF&A avoided developing the "safe" scheme that perhaps "would have been easier." Instead, they opted for the Landside Airside Concept, although, according to him "the airlines were not big fans of the scheme"; but, to their advantage, both Fisher and Brink Coridan were very close to all the airline presidents. As Lewis mentions, they would call them personally and ask "why don't you do this?" and "why don't you do that?" or "I thought about this; what do you think?" The key argument for selling the concept was that the client (the airport) "[doesn't] have to build these concourses." And with respect to the mechanized transportation, it was the also the airlines who, implicitly, "would pay for the ride system." Lewis goes deeper into the conversation and reveals one of the secrets of the negotiation:

> And the basic idea was that we don't have to build these concourses that would pay for the ride system. They [the client] could make much compact little buildings at the end, instead of all these corridors, that would pay for the ride system ... And substitute a horizontal elevator, automated, so you wouldn't increase the personnel cost. It had never been done before but it was worth thinking about. So they [the client] were just not interested in doing the easiest scheme to design. In fact, one of the planning processes was that nobody was allowed to bring a scale, a hard line drawing to a meeting because they were talking about concepts, not hard line drawings. So the engineers and the architects were prohibited, in the early stages of the planning, from bringing any hard line drawings. This is the way we did it in Philadelphia, and this is the way we did it in Kansas. We weren't allowed to do anything ... everything had to be rough, freehand, conceptual sketches. And that was one of the ways they stopped people from saying we can't do this because we don't have ten feet there, or we can't do that. (Personal interview, March 9, 2008b)

It is clear that since the early conception of the airport, the HCAA (perhaps encouraged by Leigh Fisher himself) sought large-scale innovation. However, that was only one part of the equation, because they had (as a team) to persuade those who were hesitant about the daring idea of testing a new airport scheme. Thus I find fascinating the methods that this quote reveals, and it does make me think of the strong differences between the negotiations that I have been speaking of in the previous two chapters. Gene's description of these meetings resembles more a creative think-tank or an experimental group than a meeting between experts trying to determine the feasibility of a project. However, the gestation of the landside–airside boundary has a common thread during the past decades: It needs to be set in a very conceptual environment.

6 After the "Landside Airside Concept"

In the three previous sections, I used a slightly modified version of Bijker's (1997) tentative list of elements of a technological frame. I have found along the way, particularly after analyzing the interviews, that there are three main levels to the gestation process of LF&A's "Landside Airside Concept." The first may be called "problem posing," which is strictly semantic. It is here where the relevant actors need to fix terms in order to identify complex problems and make them intelligible and manageable. Those terms may be either newly coined or existent, but new meanings must be agreed on. (Within the "posing" process, words in disuse were dusted off and put back into practice.) This is similar to the problem of music and color discussed by Wittgenstein, in which, for example, melodies may be "blue" in tone and sentiment, and thus the word "blue" does not necessarily represent or communicate the idea of the color "blue" as we see it in a clear sky. If we speak of colors, a sad, melodic "blue" may resemble more the color "gray."

Thus, the first agreement among the relevant social groups at Tampa, which is represented by actors and hierarchies during the early design process, launched a dialogue. I suggest that it is here and now that the group "posed" the problem (interestingly, no hypothesis has been formulated yet). In the case of Tampa's airport, the problem is posed at Jim

"Jonesy" Jones's basement in 1963, where the planning team *coined* the terms "landside" and "airside."[8] The internal agreement made based on the definition of those terms *is,* in itself, the problem.

The second stage I found at this atypical case may be called "problem framing" and is mainly methodological. At this stage, definitions become the main element of communication between the members of the different relevant groups. In order to achieve this sort of *rhetorical proselytism* it is necessary to outline a method or define technical procedures to take a course of action. As a procedure implies *putting into practice,* the establishment of such methodology could result in a sort of task-driven force that will become the group's *bonding material.* In the case of TIA, it is Herb Godfrey, the HCAA Executive Director, who *adopts* the Landside–Airside Concept by LF&A, and it is George Bean, his successor in the post, who finally agrees and imposes a procedure. Bean's purpose is evident; the stakes were so high with the innovative concept that the methodology had to be directed to dispel any doubts from the relevant groups. Thus the outcome was to arbitrarily fix a *comfort* standard (750ft. total walking distance) and prove all other airports in the country wrong. The simplest method, direct comparison, was the one employed.

At this point, there is a problem and a method, but no visible solution. I suggest that the third stage of this story could be called "problem solving." Assuming that not all actors are yet fully advocating either the problem or the method (or one only), planners team up one last time and, based on the procedure's outcome data, brainstorm an integrated solution to the problem. The solidity of the final plan relies thus on the following points:

- How forcefully the new layout will address the problem, which is defined by the initial terms.
- The scheme's capacity to respond to suppositions.
- The level of consensus reached, among the relevant groups.

---

[8]As I have shown in Chapter 2, this is historically imprecise. The first usage of these two terms dates back to the late 1930s, and later they fall in disuse around the 1950s. This would explain their feeling of "coining" new technical terms.

In problem solving, planners must gain enough support on all three points mentioned above in order to materialize the plan.

In the case of the airport, it is the "Recommended Planning Criteria" (or the airport scheme master plan) that will provide cohesion. Once LF&A has *proselytized* experts and sub-consultants in private, the firm is now ready to start the negotiation. The plan will be presented and discussed, or "sold," in the words of Brink Coridan. LF&A is well prepared with Leigh Fisher at the top; he had the unique ability to make the "unpalatable palatable" (Brink Coridan, 2006). In response to their interests, actors (or participants in common terms) will take their positions. How did the plan respond to the aforementioned points?

- The problem was addressed semantically and asymmetrically, assigning the characteristic of "variable" for the airside, and the feature of "invariable" to the landside. The plan is consistent in responding to the historical issues of experienced by each entity, and thus a split is proposed.
- The major supposition is that it will be technically and economically unfeasible to reconnect passengers within short times. Westinghouse sees an opportunity, competes with the less flexible Chrysler's "Mobile Lounge,"[9] and proposes the versatile "People Mover."
- High level of consensus is reached from most actors. The hardest negotiations come from the airlines, who were lobbied by HCAA's members Norm Stallings and Sam Bucklew, and eventually persuaded, thanks to the personal relationships that both Fisher and Brink Coridan wove with them through the previous years of joining the firm.

All three cases that I have addressed so far, LaGuardia, Dulles, and Tampa contain elements of Hughes's "reverse salients." The creation of a new airport model for Tampa shows how not just airport standard designs were triggering the alarms— particularly among users—but

[9]As criticized in the first volume of LF&A study, the "Mobile Lounge" was not flexible enough to cope with technological changes such as larger aircraft with higher passenger capacity or the passengers' uneasiness in case of long waits due to traffic delays.

even the so-called *revolutionary* Dulles was showing early symptoms of failure. Landside–airside boundaries are unresolved conflicts, which grow over time until they create serious bottlenecks and produce, almost cyclically, these "reverse salients." Once the airport has been *reinvented*, the protrusion disappears and the system returns to *normal mode:* in other words, back to a safe zone, where planners and designers assume standard or conventional designs, which form a sort of long, linear design progression. This continuum line acts as a stabilizing pattern. In the case of airports, this *linear development* is what later planners, architects, engineers, manager, and developers will call *conventional designs.* The classic representation of this technical idiosyncrasy is when airport designers and planners reach an apparent consensus on a *perfect airport.* This temporary rhetorical closure stabilizes and allows further developments until a new "reverse salient" prompts another questioning of the whole model. In the following years, several airports including Charles de Gaulle, Orlando, and Atlanta followed the "Landside Airside Concept." In the case of airport planning in the postwar period, atypical cases, such as Tampa, became pressure valves, in which the system was questioned from top to bottom and the critical bottlenecks of standard models were finally addressed. These pressure valves are what I have been calling here "reverse salients."

However, in spite of the "Landside Airside Concept's" influence, most plans for new airports around the world tended to return to the *standard* model in order to reduce the risks involved with the innovation. Two issues were central in rising uncertainty. First, not all airlines were happy with the idea of paying to have their own terminals—this forced them to fill out all the necessary slots, thus reducing business when not at full operation. Second, not everyone agreed with the idea of having to purchase, operate, and maintain light trains (people movers) and depend on this technology in order to function. In the end, the opening phrase of the final LF&A report of 1963 freezes on one side and predicts on the other what awaited airport planning in the next few decades: "The analyses of U.S. airport terminal buildings presented in Volume I of this report indicate conclusively that the basic reason for the tremendous size and complexity of U.S. airport terminal areas today is the conflict between landside/airside requirements" (LF&A, July

1963b). What happened after the first terrorist attacks in Israel, just three years after the opening of Tampa's airport, is where we turn next.

References

Interviews

Airport police prepared, just in case. (1971, April 11). *Tampa Tribune–Times.*

Brink Coridan, M. (2008, March 8). Transcript. *Personal interview with Marge Brink Coridan with Leigh Fisher & Associates.*

Dunlay, W. J. (2008, March 8). Transcript. *Personal interview with Bill Dunlay with Leigh Fisher & Associates.*

Lewis, E. (2008a, March 7). Transcript, Part 1. *Personal interview with Gene Lewis with Leigh Fisher & Associates.*

Lewis, E. (2008b, March 9). Transcript, Part 2. *Personal interview with Gene Lewis with Leigh Fisher & Associates.*

Sherret, A. (2008, March 8). Transcript. *Personal interview with Alistair Sherret with Leigh Fisher & Associates.*

Publications

Bane, M. (1981). *The best li'l ole airport.* Tampa, FL: Hillsboro Printing and Engraving.

Bijker, W. E. (1997). *On Bicycles, bakelites and bulbs: Toward a theory of socio-technical change.* Cambridge, MA and London, UK: MIT Press.

Bliss, A. J. (2011). *Making a Sunbelt place: Tampa, Florida, 1923–1964* (Doctoral dissertation). Retrieved from Gradworks database. (Document No. 3416656).

Brink Coridan, M. (2006, June). *Leigh Fisher associates: History and lore, as seen through the eyes of Marjorie Brink Coridan.* Publisher location and name unknown.

Culver, F. R. (1996). *Planning and implementation for the Tampa airport system replacement program.* France: AFCET.

The DC-10 will fit nicely at new terminal. (1971, April 11). *Tampa Tribune–Times.*

Douglas, D. G. (1996). *The invention of airports: A political, economic and technological history of airports in the United States: 1919–1939* (Doctoral dissertation). University of Pennsylvania, Philadelphia, PA.

Dunn, H. (1985). *Tampa: A pictorial history*. Norfolk, VA: Donning.

Those faraway places, strange-sounding names. (1971, April 11). *Tampa Tribune–Times*.

Garret, E. (1976). Designers put passengers first. *Tampa Tribune–Times*.

Getting there's easy from Tampa, Delta to share airside no. 3. (1971, April 11). *Tampa Tribune–Times*.

Godwin Brown, K. (1994, June 6). World-famous TIA spawned from former WWII air base: Know your Tampa, a weekly history lesson. *North Tampa*.

Gorham, H. (1971a, April 11). Tampans dazzled by their airport. *Tampa Tribune*. New and Features, Section B.

Gorham, H. (1971b, April 11). They came, they saw, they liked. *Tampa Tribune–Times*.

Gorham, H. (1971c, April 12). New airport jammed on open house: They came in greater thousands. *Tampa Tribune*.

The greening of an airport. (1971). *Tampa Tribune–Times*.

Hailey, A. (1968). *Airport*. Mishawaka, IN: Doubleday & Company.

Hawes, L. (1978). Drew Field was a world unto itself. *Tampa Tribune*.

Hill, J. (1996, April 14). Tampa airport looks good for its years. *Tampa Tribune*.

Hillsborough County Aviation Authority. (1965). *A new approach to jet age air terminal development: Progress report 1961–65*. Tampa, FL: Author.

Huettel, S. (1995, September 9). Time keeps on moving: Tampa International Airport will retire some of the first shuttles ever installed at an airport this week. *Tampa Tribune*.

Hughes, T. P. (1983a). *The evolution of large technological systems: The social construction of technological systems*. Cambridge, MA: MIT Press.

An impression: The city will have to keep up with its airport. (1971, April 11). *Tampa Tribune–Times*.

Leigh Fisher Associates, Inc. (1963a, April). *Study of U.S. airport terminal building as a basis for planning new passenger terminal facilities at Tampa International Airport* (Vol. 1). San Francisco, CA: Author.

Leigh Fisher Associates, Inc. (1963b, July). *Recommended planning criteria for new passenger terminal facilities at Tampa International Airport* (Vol. 2). San Francisco, CA: Author.

New 180-foot control tower due, to cost $1.4 million. (1971, April 11). *Tampa Tribune–Times*.

Nordle, P., National Aviation Facilities Experimental Center, Human Sciences Research, Inc., et al. (1961). *Airport transportation: A study of transportation means between airports and the metropolitan areas they serve*. Arlington, VA: Human Sciences Research.

O'Donnell, B. (1991, May 14). Designs readied for TIA airside. *Tribune*.

Pinch, T., & Bijker, W. (1987). The social construction of facts and artifacts: Or how the sociology of science and technology might benefit each other. In W. E. Bijker, T. P. Hughes, & T. J. Pinch (Eds.), *The social construction of technological systems*. Cambridge, MA: MIT Press.

Rane, J. (2011, November 17). Web poll results: 10 of the world's most loved airports. *CNN Travel*. Retrieved from http://travel.cnn.com/explorations/life/10-most-loved-airports-981939.

A Santa in April, officially speaking, smiles are in order. (1971, April 11). *Tampa Tribune–Times*.

State's skies getting congested as volume of air traffic grows. (1989, June 5). *Florida Times–Union*.

Tampa International Airport. (1972). [Brochure on Tampa parking system and terminal].

Tampa's new crown jewel. (1971, April 11). *Tampa Tribune–Times*.

Taylor, E. (1986, June 29). Airport wins praise from airlines. *Tampa Tribune–Times*.

Terminal sprawl, no ailment here. (1971, April 11). *Tampa Tribune–Times*.

Waresh, J. (1987, April 20). Bean helped Tampa airport win its wings. *Tampa Tribune*.

Westinghouse Company. (1971). [Illustrated advertisement insert]. *Times*. Tampa.

Wittgenstein, L. (1970). *Zettel*. Berkeley, CA: University of California Press.

5

"Are Landside–Airside Boundaries Cultural Mirrors?": Reinventions, Innovations and Society

Are boundaries standard pieces of a mechanism, or are they autonomous entities, subject to absorbing cultural practices? Let us keep in mind that for Heidegger the essence of technology is by no means anything technological. Thus, this chapter will explore the external agents, particularly the social ones, that give meaning to the airport as a technological artifact. But before I tackle this question in the first half of the chapter, I would like to draw the line between airport reinventions and airport innovations and thus help the reader by dispelling any blurriness on this matter created by the previous case studies. In the second half of this chapter, will I argue why these boundaries are capable of reflecting the spirit of a specific society (Fig. 1).

More widely, in this chapter I argue that airports are not clones or even repeatable copies. I support this claim by showing that the frontier between the airside and the landside captures many cultural differences that single out each airport through history. These singularities produce a mirror effect that reflects either local customs or the way these local forces react to foreign or global models. Toward this end, I speak of three topics: causes and legacy of airport reinventions, the process of landside–airside boundary innovation, and how these boundaries add transparency and allow us to glance at the cultural mirror.

V. Marquez, *Landside | Airside*, https://doi.org/10.1007/978-981-13-3362-0_5

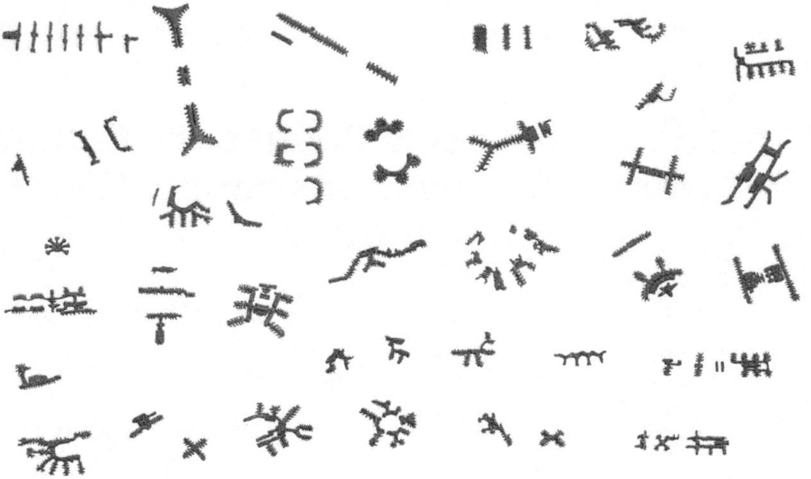

Fig. 1 An artist's representation of airport morphology, showing the diversity of forms within this building typology. The world's 30 busiest international airports are presented as terra cotta bas-reliefs, hand-built to the same scale and all oriented to True North. Abrams, Janet [Pieces in terra cotta. Installation: 10 ft × 16 ft] *In the Unlikely Event* (2013) (Reproduced with permission of Janet Abrams [www.janetabrams.com])

In Sect. 1, "On Airport Reinventions," I discuss how congestion can be highlighted among the causes that prompted full airport reinventions. I also comment on the consequences of my previous three case studies: LaGuardia, Dulles, and Tampa. Section 2, "On Airport Innovations," speaks of historical changes, departing from what is commonly known in airport planning as the "linear model." I present two iconic cases that introduced airport layout ideas that were perceived as new: the 1974 Terminal 1 in Paris's Charles de Gaulle Airport, and the 1960 Pan Am Terminal 2 in New York's Idlewild Airport (later known as JFK). Last, in Sect. 3, or "The Cultural Mirrors," starting off from the notion that all airports seem to be different, I argue that landside–airside boundaries may reflect cultural practices and customs and show how that led me to maintain that airports, as technological systems, are socially constructed. In addition, I address the differences between planning new airports and expanding them in phases, and, finally, I compare boundaries at regional and international airports.

1 On Airport Reinventions

Sociologist Everett Rogers defined technological reinvention in 1962 as "the degree to which an innovation is changed or modified by user in the process of its adoption and implementation…re-invention as the degree to which an individual's use of a new idea departs from the machine version of the innovation that was originally promoted by a change agency" (p. 17). Thus, I argue that to set the right conditions for reinventing technologies, first there should be agents of change that push innovators to learn from the previous model and launch them into the task of redefining a whole new concept for that artifact, product, or design. In the previous chapters, I have introduced these agents of change in each of my case studies.

Likewise, if we use Hughes's (1987, p. 62) approach, for him inventors can be divided into conservative or radical. A radical invention is one that inaugurates a system, closer to the full breakthrough of the cases that I have presented. Hughes suggests that inventors often speak of their inventions as metaphors or analogies, as with the Skywalk, the Mobile Lounge, or the People Mover.

Hughes also predicts that preexistent and new systems will enter into competition. So how can we evaluate these three case studies in competitive terms? There are two contrasting ways to answer this question. On the one hand, all three cases tell stories of technological failure because they did not bring a definitive closure to the problems that originated them. In the case of LaGuardia Airport, the planned capacity reached its peak just a few years after its opening, making the investment obsolete to the point of demolition. Unfortunately, demand at the terminal and room for expansion were seriously underestimated. Plans to consider a new airport for New York were undertaken just a year after the airport's opening and even during the Second World War, leaving aside the ideas of William Delano and embracing William K. Harrison's futuristic and pragmatic vision of planning an airport city, rather than just another airport. In the case of Dulles Airport, despite the fact that the new airport system designed by Saarinen, Chrysler, and the FAA was widely promoted and adopted worldwide, airports gradually decided to return to the old system of jetways and were forced to send their expensive and bulky Mobile Lounges to the junkyard. Because of some minor miscalculations in the system's model, Dulles's scheme could not cope with the pressures of growth. Planes at the huge apron were subject to delays and overlapping, so in the absence of large hold rooms, the terminal building became saturated and passengers were forced to wait for long periods on board both planes and Mobile Lounges, making the aircraft turnaround service slow and inefficient. Last, Tampa's airport responded to an ownership model, according to which airlines developed their own terminals at their own cost, that did not last for long. In addition, the layout's radial geometry obliged the construction of oversized aprons, which required at the same time enormous extensions of land. This did not scale well to smaller airports that were looking for less-expensive models for expansion. Last, Tampa fell at the opposite extreme of Saarinen's TWA terminal in New York and Yamasaki's Lambert Terminal in Saint Louis: Florida's terminal lacked any monumentality or architectural taste, making it one of the all-time least favorite public buildings in Tampa Bay.

On the other hand, few airports have ever been more influential than the ones studied here. As I commented earlier, LaGuardia made the significant contribution of separating not just passenger flows, but the entire airport activity in vertical levels. This has remained the norm throughout the years, particularly for international airports, and certainly marked a radical new way to deal with the landside–airside relationship. Separating arrival and departure passengers proved to be a good way to speed up drop-offs and pickups at the curbside; inside the building, it facilitated the handling and recovery of luggage, preventing people from flowing in opposite directions through narrow spaces. It also simplified the way passengers were distributed toward the airside, allowing for shorter, roofed walkways at the apron. Even the Skywalk was listed in the requirement program of other airports or expansions, such as Washington National Airport (1941) and Zurich International Airport (1946).

In the case of Dulles, Saarinen proved that aviation is a business plagued with externalities, so there is a high level of risk if an airport is turned into a precision clockwork. However, Dulles's Mobile Lounge became an instant option for already congested airports in the 1979s such as JFK, Los Angeles, Philadelphia, Montreal, Tokyo, Mexico City, Jeddah, and many others that wanted to increase the number of operations at remote positions—without making many changes to their existing buildings and sacrificing much in terms of passenger comfort. This way of adding capacity through shared, insufficient hold rooms provoked a direct relationship between delay and incredibly crowed spaces. Thus, designers also learned that it is very hard to uncouple the number of enplaned passengers from the dimensioning of the terminal buildings. As every story has its moral, many officials learned that experimentation in airport design comes at an incredibly high cost.

Finally, of the three reinventions of the landside–airside, Marge Brink's vision for Tampa has been the most influential. Why? Because it founded the basis first for Orlando and later for Atlanta international airports, and because up until the time of this writing, only Georgia's capital airport is thought to be a viable model for growing massive, hub-like international facilities. What many of Marge's talented disciples did was to replace the spiked like layout with a "tablet"-like arrangement of conventional linear terminals linked by rapid ground transport.

So we may consider that Tampa added one of the key features of new hub—airport planning: the mechanized connection between satellite terminals. Among the facilities that have embraced and adapted this model are Stansted in London, Denver in Colorado, and the forthcoming London Oxford Airport in the UK and, paradoxically, the recent SOM's[1] re-conception of Dulles Airport in Washington.

The planning legacy of these three case studies is rather unusual. None can be necessarily described as historical turning points because their importance was right at the core of how landside–airside boundaries were conceived, and as I have argued in this book, the boundaries were not even verbalized as planning conflicts until Brink Coridan did so. Although they certainly have had an impact, at least in the way the

[1]Skidmore, Owings and Merrill, which, since the postwar period, has consolidated into one of the largest architecture and engineering firms in the world. Its airport division has been led by two key individuals: Airport Planner Tony Vaccione and Airport Designer Marilyn Jordan Taylor.

history of airports has developed, planners still have not agreed on how to deal with the airport as an entity of constant and onerous change.

One last aspect that should be addressed is the common cause that triggered a radical reinvention of the airport. All three cases, LaGuardia, Dulles, and Tampa, are justified first and foremost as countermeasures in response to congestion. This phenomenon is sensitive to favorable economic conditions where taxpayers are more prone to travel and spend, but inversely proportional to boosting airport income and revenue, facilities become rapidly overcrowded and insufficient.

These issues are intricately linked, from the airspace slot availability to the lack of parking or curbside space, thus creating an incredibly complex maze of problems when capacity becomes swamped. In the aforementioned cases, only politicians and institutions (such as Mayor LaGuardia, the FAA, or the HCAA) were hierarchically capable of driving a total reconceptualization of the airport.

Fortunately, not all airports need to be reinvented when they face a saturation crisis; however, most airports need to implement their own program of innovation (or at least preferably) in order to cope with potential growth. I discuss this aspect at length in the following section. First, however, I explain why congestion has become one of the strongest forces driving the continuous change of the landside–airside boundary.

As *through spaces*, transportation terminals should facilitate the flow of passengers. If, according to International Air Transport Association (IATA), "the essence of air transport is speed, then it is essential that delays on the ground are avoided. Passengers who are subjected to congestion, inconvenience or delay tend to blame the carrier for any shortcomings …" (as quoted in Brancker, 1977, p. 22). This congestion must to be predicted based on the previous experience of consultants and airport administrators and thus will inform the dimensional-requirement program of airport facilities, in order to guarantee minimum speed patterns. As a result of market fluctuations, airports are subject to handling an increased number of planes and passengers, which often overwhelms the capacity of their infrastructure.

Technically, saturation is measured in peaks and delays; these figures usually feed back the minimum standards written in codes and the so-called levels of service (LOS). These minimum levels of comfort are set by IATA and are dictated by the maximum allowable human density per square meter. It is measured on a scale of letters, ranging from a high LOS of A for comfortable 1.4 sqm per person to a low LOS of E for 0.6 sqm per person, for example, in customs queuing (Poh, 2007, pp. 16–18). Thus, much airport planning activity revolves around the application of analytical tools and methods to construct prediction models. Besides calculating growth volumes, forecasting may prevent airport jams in the future. However, the reasons we travel are impossible to predict with high accuracy; society as a whole becomes an externality in the model. When passengers are affected, they usually externalize their criticisms and complaints. Historically, congestion has reached several high peaks that eventually led to management crises.

Some of the most famous critical peaks in history are the saturation levels in the late 1930s—most probably caused by the gradual recovery from the Great Depression and the generalized excitement of flying; the explosion of commercial air traffic and airliner development after the Second World War from 1945 on; and a new striking peak unleashed with the introduction of jetliners in 1954 wherein travel times and costs were significantly reduced. Airport congestion became a serious problem during the 1960s when, after the general airline de-regulation, a sort of war for aerial slots was triggered (Levine, 2008, pp. 56–58). However, a closer look at saturation shows that it is not necessarily linked to technological innovations; for example, comparative statistics on travel from the 1960s until the 1990s reveal passengers' motivation for personal travel: "In the 1970s the specific reason that showed the most dramatic gain was sightseeing and resort use, followed by people flying to visit friends or relatives" (Goodman, 2000, p. 28).

According to Goodman's (2000) report, leisure travel grew faster than business travel during those years; flying for personal travel purposes, even internationally, became increasingly popular. Congestion detonated debates over the technical and managerial capacity of airport experts and authorities to deal with saturation and the consequential lack of comfort for users. This has moved the landside–airside boundary into the arena of discussion. As congestion implies delays—thus making passengers experience longer waits—assigned spaces for holding, eating, shopping, or service areas such as lavatories or circulations become physically insufficient. Congestion has directed pressure toward the boundary in very different historical contexts, as the limits of control and surveillance have changed throughout the years, preventing passengers from wandering around terminals and, on the contrary, forcing them to remain in contained zones. As I discuss later, overcrowded hold rooms became fish tanks of anxiety; thus planners, designers, and managers were forced to attend to this matter and address these issues at the negotiation table.

2 On Airport Innovations

In this section, I first sketch a very brief history of the "linear airport," which, as the reader may infer, is the simplest way to grow an airport terminal. In this sense, I also suggest that most innovations sprouted from a problem-solving conjunction related to this simple, more standard design. Further ahead, I interlink the basics of innovation theory with airport changes, and I draw the differences between airport innovations and reinventions. Last, I analyze in more depth two cases of paradigmatic landside–airside boundary innovation. So, what went wrong with the so-called linear scheme?

Let us imagine first a huge rail terminal station, where trains that are more than one mile long—four times the standard—are standing at a hypothetically long platform, ready to be boarded. This length would be comparable to twelve to sixteen city blocks. Most likely passengers would not feel very happy to step on in at the few last cars, especially after a long walk sliding their luggage and shopping bags. This imaginary analogy captures one of the most basic landside–airside conflicts: airplanes requiring more "building front" versus passengers hating the long walks. Perhaps Callon or Latour would have claimed here that aircraft, just as many other technologies, have performed some degree of *actorship*.[2] True, planes just as other artifacts give rise to physical

[2]Referring to S&TS's controversial Actor-Network Theory (1986), wherein inanimate objects may be "actants" or things made to act and thus present some degree of *actorship*.

demands; however, all these demands have been pushed ahead by human means and thus can hardly be considered autonomous.

After many years of UK and German predominance in modeling early symmetrical Art Deco terminals such as Liverpool Speke (1938), Dublin (1937), or the more functionalist Hamburg-Fühlsbuttel (1928) and Stuttgart-Echterdingen (1939), the French became the strongest promoters of the linear airport. After the emblematic extension of the old Le Bourget airfield in Paris by George Labro in 1937, the rationale of linear schemes prevailed. The French architect responded to aircraft demands and imagined one long building, centrally divided by a core, expandable at both ends. Although the structure ended up being a quarter of a mile long, it became a paradigm and thus the more common way to grow existent airports worldwide even after the Second World War. *Linearity* remained a preferred option—even during the 1960s' and mid-1970s' intense period of airport questioning and innovation, mainly because following the steps of Dulles, Tampa, or Charles de Gaulle implied prohibitive costs and investments. (Remember that in addition to the new terminals and transportation systems, they all required massive aprons, which came at a high cost.)

But even Le Bourget's terminal linearity was no match for the more radical, 1945 longitudinal scheme by Paul Gerhardt, for Chicago's Midway Airport. The visionary city's architect proposed a lengthier low-cost terminal, capable of parking up to fifteen planes in front of the building's façade. Surprisingly, the plan remained unbuilt.

Just when the first professional airport planners were trying to revolutionize layouts and systems, Henri Vicariot, an architect-engineer from *Ponts et Chaussées*, designed for ADP the ultramodern, legendary, and Miesiean-like[3] Paris Orly International Airport. The flashy 1971 terminal privileged aircraft, as it was disposed in a long (660 feet), rectilinear way, parallel with both the apron and the runway, thus facilitating the perpendicular parking of planes. The terminal certainly captured the spirit of modern architecture, as it was dressed under an austere glass skin layered out by flat, concrete slabs. Orly instantly became a

[3]*Miesian* is a term often used by architecture historians, when referring to the bare, unadorned, distinctive style of German architect Ludwig Mies van der Rohe.

cultural fetish, dragging endless number of artists and filmmakers to its grounds, seeking futuristic and abstract backdrops (e.g., the airport is superbly portrait in the cult-classic film *La Jetée*, from Chris Marker). Gradually, linear airports became an easy way to cope with an increasing demand for simultaneously servicing multiple aircraft, although this convenience came with a price. When an airport terminal's footprints extend along a line—be it buildings or piers—passengers' walking distances will increase proportionally, but so will technical systems and, in particular, segregated circulations. Hence, since the 1970s designers and engineers have developed mechanized alternatives, such as people movers and automatic luggage systems, which help to alleviate the transport of passengers, supplies, and bags along extended distances.

It is pertinent to clarify first that most authors refer to "linear" terminals as straight buildings. From an architectural point of view, this usage is correct; however, in terms of planning, the *linear terminals* are the ones parking and servicing aircraft in a line or following a perpendicular angle to the building. That would imply that linearity is a way to refer a landside–airside condition rather than an architectural shape. Therefore, airports following a star footprint, such as Amsterdam Schiphol, or even curved, as Dallas-Fort Worth, could be well considered *linear*. This condition changes when we speak of satellites—which were very much in fashion between the sixties and the seventies—because the relationship between the airside and the landside has been strongly altered or rephrased.

But why is the airport's *linear scheme* still used at present? The scheme offers two advantages: First, it privileges aircraft, making the parking simpler and the aprons shorter, and second, it is easily expandable—if the land provision suffices—explaining why this is the classic solution for new airports and terminals. Iconic examples of contemporary rectilinear terminals include the new Terminals T4 and T4S at Barajas, Madrid, Terminal 2 at Detroit, Michigan; Kansai International Airport in Osaka, Japan; and the huge Dubai International Airport in UE, underway at the time of this writing. These and all linear projects must deal in creative ways with the oldest landside–airside conflict: the walking distance.

The aforementioned examples also share the use of mechanized systems in order to mitigate the long walks; for this purpose, Kansai implemented the "Wing-Shuttle," Detroit the "Express Tram," and Barajas its "People Mover." Today, more than 40 airports use similar or identical models to the ones pioneered in Tampa, as I discussed in Chapter 4. In time, the "linear scheme" became a synonym for passenger discomfort and a professional target for airport planners and designers. In my view, this was one of the main reasons to produce relevant, airport layout innovations throughout the twentieth century.

According to the influential sociologist Everett Rogers (1995), "an innovation is an idea, practice or project that is perceived as new by an individual or other unit of adoption" (p. 14). Because of the complexity of airports, innovations are required far in anticipation of problems such as forecasted congestion, user complains, airline demands, or technological advancements in aircraft. Therefore, airport planning may involve high degrees of technological speculation and design, which occasionally result in innovative landside–airside configurations. I suggest that these changes exclude reinventions, because these last respond to a different decision-making process, whereby a huge deal of uncertainty is paired with an enormous cost. On the contrary, airport innovations seem to make good use of past solutions and schemes, thus introducing novelty in limited degrees, hence proportionally reducing risk. Among the numerous innovations that have been registered through airport history, two stand out as the more extreme and revolutionary. First, I will speak of the 1960 Pan Am Terminal 2 at New York's Idlewild Airport (later known as JFK) and later of the 1974 Terminal 1 in Paris's Charles de Gaulle Airport. Consistent with Rogers's theory, landside–airside boundary innovations show a rate of adoption and a pattern of diffusion. In the case of Idlewild, the new model has been successfully adopted and adapted throughout the years worldwide; on the contrary, those who followed T1 at Charles de Gaulle have faced serious problems coping with both technological and social change. The following two examples are also representative of two different approaches: In the case of New York, the scheme is clearly open and adjustable; in the case of Paris, however, the layout is only repeatable but hardly expandable or flexible enough to respond to change. Next, I describe in more detail the case of Idlewild, or JFK, as it is better known today.

During the formative days, say before 1937—when the extension of Paris Le Bourget was completed and the first linear terminal was born—all airports were mostly speculating or swimming in untested waters. A glimpse at the airport designs before this date may evidence how "architects were absent from the field of airport development," as airport historian Geza Szurovy (2003) has argued and that airports "continued to be an engineering problem—a viewpoint jealously guarded by the engineering community. Air terminals were expensive and, in the view of practical engineers who care little for marketing and customer service, added little value to the airport" (p. 57). In fact, both engineers and architects were pioneering what airports could be, and thus, it is undeniable that everyone involved learned much through trial and error. In this sense, by general consensus Berlin's Tempelhof arises as the greatest of its era,[4] although, paradoxically, it did not became a model for future airports. The relationship between its landside and its airside was perhaps the perfect fit for the era and the German culture. There was also the issue of the airport's own DNA-duality, or a dilemma between being thought of as a civil or military facility. Not surprisingly, much of the landside–airside boundary layout and protocols at Berlin's capital airport were designed specifically to accommodate the Führer's requirements for hierarchy, security, circulation, and privileges of observation of the tarmac.[5]

Thus, I believe it is relevant to ask whether, beyond the case studies addressed in the previous three chapters, there is record of other cases of landside–airside boundary innovations after the Second World War. The answer is yes, because almost each boundary has been custom-designed since, so in a way, each is unique. However, this fact does not necessarily imply that each of these thousands of airports worldwide can offer technical innovations that may be extended to other airports and create a wave of influence. Nonetheless, I would highlight at least three additional cases that have proved to exercise a strong impact on other similar conceptions, all of them addressing landside–airside boundaries as their *leitmotif.*

[4]A system of circulations that would lead Hitler to his private observation balcony.

[5]Direct observation in a visit to Tempelhof, guided by the Berliner Flughäfen.

The first interesting case sprouted directly from LaGuardia Airport's immediate saturation and eventual failure after investing $45 million in public funds; New York City required a much larger airport, and this decision now fell under the control of New York Department of Transportation (NYDOT) instead of the Mayor's Office (thus leaving William Delano out of the contending). For NYDOT, *rapid obsolescence* was totally out of the picture, so after disregarding a 55-gate linear proposal and after expropriating land for over 150 million dollars, it was finally Wallace K. Harrison—an architect who enjoyed a close relationship with the influential Nelson Rockefeller—who became responsible for developing Idlewild's final master plan at the turn of the 1950s.

Harrison got the job done with great vision and accuracy, proposing what he called an "airport city" or, in economic terms, a new privatization model for the larger-scale airport. In this model, public funds would be responsible for financing just the landside facilities, and airlines would have to pay for their own airside buildings.

His airport city consisted of a central curved-shaped area holding parking bays and public spaces, edged outside by a multi-lane road, which at the same time organized many independent terminals sprouting perpendicularly from the center. According to this scheme, each terminal would be assigned to an airline, and airlines would be free to invest according to their own predictions of growth. As airlines have predicted, the air boom was unstoppable. Harrison's innovation thus was to see through the landside–airside boundary an opportunity for new ownership and investment at a never-seen-before scale, which may now be considered the pioneer of today's hub airports.

A decade later, and curiously at the same Idlewild, Walther Prokosch, co-author of the first airport planning manual (Prokosch & Froesch, 1946) and one of the first airport planners in the USA, was charged with designing—along with his firm Tippets Abbet McCarthy Stratton and architects Ives, Turano and Gardner—the so-called *Worldport*, Pan Am's iconic Terminal 2. This complex was planned as a multilevel triangle facing the road and crowned, at the opposite end, with a huge, oval roof facing the apron. Similar to the much older Tempelhof, the roof was in fact an overhang, which had become unnecessary since the growing use of jetways after 1959.

But the *Worldport* terminal was meaningful as a landside–airside innovation in one specific way: It introduced a failed, but provocative, way to reduce the distance for uploading and downloading passengers. The first triangle had a parking garage at the rooftop, and one flight below, the hold rooms connected to six contact positions. But the terminal designers literally chopped off whole sections of this building in order to allow each aircraft to insert its nose into the landside (at least at the first building). This radical move showed a strong concern for and commitment to reducing boarding distances and to increasing passenger comfort, as emplaning required only short bridges.

Unfortunately, the final building did not feature this system, probably because it was not compliant with the minimum distances recommended in *Annex 14* regarding the proximity of aircraft to built structures. It also required push-back from each aircraft, making it additionally slow for service turnaround. The learned lesson of the Pan Am terminal is that it is not advisable to overlap the airside with the landside. This principle remains valid, at least until some futuristic plans that will be discussed at the end of this chapter.

The last innovation that I would like to bring forth is the first terminal at Roissy in Paris, later known as Charles de Gaulle International Airport. This airport is closely linked to the case of Tampa's airport, as there are some striking parallelisms in the thinking processes; thus, I will explain further—although the Parisian terminal is perhaps better known and more iconic—why Tampa International remains the last reinvention of the airport.

Before the evident saturation of Orly and Le Bourget in the early 1960s, and after years of negotiations to build a new airport in the surroundings of Paris, it was Jacques Block, a director of planning and development for *Aéroports de Paris* (ADP), who was put in charge of the master planning of Roissy. However, the French administration was caught in the same conflicts as those at the FAA in the USA, dealing with the endless passenger complaints about having to walk long distances and bear long waits at congested concourses. Thus, before developing the airport zoning layouts, Block decided to challenge the newer airport layouts in the USA and set forth only the design of the first terminal (as if he were prototyping or experimenting). In the meantime, he would have time to negotiate the land provision for the larger airport plan, which needed to be pushed twenty-five kilometers out from the old city center (Block, 1971). In 1968, during a lecture at the Royal Aeronautics Society in London, Block assured that his model for "airport planning was influenced by aircraft developments," thus setting the tone for the new designs. ADP denounced the American model of Dulles Airport, arguing that the revolutionary scheme failed to continuously feed passengers to aircraft, thus limiting its capacity. They also challenged the linear model at Orly and the recent addition of piers, or fingers, because they implied long walks for passengers.

Finally, ADP showed interest in the first satellite terminals in Newark, although they (Block, 1968) argued against distances and a limited number of positions (Aéroports de Paris, 2004). As Jean-François Onnée wrote in his clever analysis of Roissy's Terminal 1, "it was meant to reflect the French 'savoir-faire' in civil and aeronautical engineering" (Onnée, 2004, p. 7). Within ADP, it was Paul Andreu, a graduate architect and engineer from both the *École Polytechnique* and the *École Nationale des Pons et Chaussés*, who led the project as chief engineer of the company. In 1967, Andreu envisioned a multilevel, circular building at the center containing the parking and processing of passengers; this drum-like structure was surrounded by a huge circular air-platform, equipped with seven self-standing mini-terminals, radially spaced around the core. Tunnels and people movers physically linked both the satellites and the center, thus allowing the flow of passengers and luggage.

Similar to Tampa but unlike any other airport, Roissy's Terminal 1 concentrated parking and shops on the first levels (1–3), followed by the check-in hall level (3), then an intermediate connecting level toward the satellites (4), and finally topped by the arrivals hall (5). The building was completed and unveiled in 1974 (Aéroports de Paris, 2009). In an article titled "Charles de Gaulle Terminal 1: The Hated Air-Terminal," the weekly *L'éclat* noted that at Roissy's first terminal, "the lacking of character had largely contributed to the disaffection of the public." The editorial added that "the poor choice of materials created a bad reputation among users" and finally declared it a "failed utopia" (L' ECLAT rédaction, 2011, pp. 2, 4).

In practical terms, the layout was so rigid that it did not allow for any expansion or adaptation, making it very uncomfortable especially after the rise of security controls. This gave the terminal a bad reputation, making it, through the years, highly unpopular among travelers. The unusual scheme certainly challenged its users. In his layout, Andreu only "extended" the frontier between the landside and the airside, which would explain why this division was never verbalized as a rupture.

On the one hand, Terminal 1 is, like Dulles, a building designed primarily by architects; aesthetically radical and spatially sound, it has been displayed in films and photographs as futuristic.[6] (Who could forget the collage of Terminal 1's atrium and its escalator tubes on the awesome cover of the 1977 *I Robot* album by the Alan Parsons Project!) On the other hand, there were too many flaws and contradictions in the way passengers used the building, and as Onnée (2004) criticized, despite its centralized vision,

> ADP's concerns was [*sic*] to keep the walking distances for passengers as short as possible. The finger pier concept was criticized for its inefficiency in that respect. But this design did not actually prove to be efficient either. The overall walking distance hence turns out to be not negligible at all, not to mention the fact that passengers are constantly being disoriented and deceived by the circularity of the terminal and the fact that the different levels do not connect in an obvious way. (Onnée, 2004, pp. 10–11).

One more substantial difference between the two airports was the ownership model. Roissy remained a centralized facility owned by the state (through ADP), in contrast to Tampa's model of independent *satellite* ownership per airline.

Nonetheless, for the purposes of this book, I analyzed both projects not from an architectural historian perspective, but under the light of S&TS scholarship and through the eyes of the landside–airside

[6] As described in a 1974 journal video from ORTF, *L'installation d'Air France dans le nouvel aéroport Roissy-Charles-de-Gaulle.*

boundary. In this regard, I conclude that the main difference between Terminal 1 at Roissy and Tampa's airport is that the first represents above all a meaningful architectural achievement and the second a *socio-technical* reinvention of the airport. In Paris, the boundary has been stretched out to a maximum, while always leaving spatial continuity between both landside and airside. On the contrary, Tampa's scheme splits the boundary in two, with landside on one side of a fence and airside on the other. The new relationship would be *technologically negotiated* by a machine (People Mover). This is what phrased the relationship in a new way, and what eventually served as the basis for Atlanta and other contemporary, large-scale airports. In short, the key to airport future expansion was Tampa's divide.

In conclusion, in the last two sections I drew the line between two kinds of airport layouts—or, more specifically, landside–airside boundaries: innovations and reinventions. I also commented on what lies behind these models or, in other words, factors that motivate change such as congestion, the "linear airport," and passenger discomfort. Finally, I used two compelling examples of airport innovation to illustrate how these new layouts may respond to external forces through time, evidencing thus how open or closed these systems are. Now I move on to analyze how local culture influences the formation of any landside–airside boundary.

3 The Cultural Mirror

Even if airports seem to be similar or share certain common features, landside–airside boundaries are strictly unrepeatable. For example, American artist Janet Abrams shows with great sensibility, that is, all airports are morphologically different. In her striking piece *In the Unlikely Event*, Abrams represents, in the format of a large-scale installation, unique terra cotta pieces of individual terminals that compose something that looks like a jewel box. Thus, airports and boundaries are expanded typologically into a sort of endless diversity. But yet in cases where designers tried to impose identical models at two different terminals, those boundaries would eventually be customized and individualized. This is the case of near-to-identical terminals in the Los Angeles and Dallas–Fort Worth airports, which, depending on a number of factors, may substantially vary as an individualized boundary. For example, in the case of Los Angeles, terminals handling flights bound for Australia or Asia have recently adopted zigzag ramps within a small area in order to process the very long lines produced by embarking hundreds of passengers to the oversized 747s. Also, in the case of Dallas–Fort Worth, where all five terminals are deceptively identical, statistics show that out of nineteen inspection points distributed along the five buildings, wait times may vary up to twenty-one minutes between the fastest and the slowest.[7]

The landside–airside boundaries at Dallas–Fort Worth are contingent on how old the terminal is (A is oldest, D is newest)—because there are may be differences for each in their interior designs, the kind of equipment being used, the operating airlines, destinations and kinds of aircraft, the domestic or international nature of each wing, retail brands, security checkpoint layouts, and in particular the personal stamp of administrators, officials, employees, and users. Surprisingly, even in this type of airports, no two boundaries are the same.

So, why are all landside–airside boundaries different? I suggest that there are at least three main reasons to make each landside–airside frontier singular. First, there is a contextual (physical, political, managerial, economical) difference between planning new airports and expanding existing terminals. Second, there are two basic kinds of airport boundaries: domestic frontiers and international borderlines. Third, landside–airside boundaries have become representations of cultural expressions.

The main difference between planning new airports and expanding existing terminals is the degree of freedom on the hands of designers, architects, and planners to come up with solutions that respond to the specific needs of an airport master plan. In the case of expansions, designers must adjust existing terminal buildings and aprons, which

[7]"Dallas DFW Security Wait Times," TSA historical averages reported in 2010, retrieved from www.ifly.com.

are often seriously constrained in terms of room for modifications and additions, and are typically congested.

If the activity of negotiating, designing, and implementing changes to older terminals is not complicated enough, decision makers must also deal with preserving a continuous operation scheme. In airport planning, this is usually referred as phasing. A representative example of an airport expansion that became incompatible with the airport's landside–airside scheme could be Tampa: After "the attacks in Rome and Athens in 1973, the American Federal Aviation Authority (FAA) imposed new rule narrowing the points of control that were established at the "throat" of each concourse" (Salter, 2008, p. 13). Bill Dunlay, an engineer with Leigh Fisher since the 1970s, remembers how the rise of security immediately affected Tampa's Landside–Airside Concept, particularly on the first airports that followed the same scheme:

> Security was at the hold room right before going into the plane. That's how airports were built long before security. When Orlando was first open (same as in Tampa), I believe security was on the airside building, but later security was a known function. Everywhere else the security is in the landside building, rather than the airside. So it became a major factor in the United States, for that function that didn't exist before and hadn't been planned for before. (Personal interview, Bill Dunlay, LF&A, March 8, 2008)

At both Tampa's airport and its predecessor, Orlando International Airport, there was simply no available space to install landside security checkpoints. The functional scheme was fully locked up with Leigh Fisher's "Landside Airside Concept," the connectors between the airside satellites and the landside central building ended up on a small, mezzanine level. Furthermore, after the events of 9/11 this situation was pushed to the limit; airport authorities had no option but to protrude oversized security checkpoints and their queuing lines into the smaller vestibules at the transfer level, resulting in an overcrowded, uncomfortable area for passengers. Paraphrasing Slater, if, in general, boundaries were pushed to the terminal's *throat*, in the case of these two airports it almost choked them.

In contrast, when designers have the chance to lay out new airports from scratch, the task of shaping landside–airside boundaries turns into a rather complicated negotiation under a rain of ideas, interests, and restrictions. While additions or upgrades to already-built terminals face several constraints, these same limitations act as negotiation plotlines, in which there is less room for imagination. But in the case of laying out new airports or terminals, the possibilities are certainly expansive.

I have participated in such negotiations from at least three different angles: First, as a designer for Toronto's International Airport in the late 1990s; second, as an architect and project leader for Monterrey's International Airport from 2006 and on; and, finally, as planner for Mexico City's Toluca International Airport since 2012. As one might expect, these three projects were conceived, developed, and decided on the basis of very different negotiation platforms. Why? Perhaps because for each decision's group, there was a rather different leadership profile, often representing highly opposing interests and ideas. In addition, there were a number of concerns, clearly representative of their time, flying around all the time.

In the case of Lester B. Pearson in Toronto, the more heated debates spin around the design of an efficient segregation system—airport officials were worried about the USA, Canada, and all other nationalities' triple separation. I recall that the long discussions between us, the designers (Moshe Safdie and Associates), and SOM, our airport expert partner, ended up in a project split: We would be in charge of designing the piers, and they would be in charge of defining the boundary, at the processor building. The outcome was an intricate bundle of segregated *tubes and valves* distributed into huge, monumental processing spaces for passengers. We aimed to guide passengers through light patterns and display artistically designed structures; however, a good deal of the final layout relied on complying with those migratory conditions that would allow the airport to compete commercially with its American peers. I also recall Miron Cohen (a project manager for Safdie's Ben Gurion expansion project), who was brought from Israel to reinforce our young team, mentioning for the first time the expression *fish tanks* to refer to rooms where passengers remained isolated from others; Michael Guran, the project associate, became simply obsessed with producing endless zoning diagrams that could solve the puzzle-like challenge of the maddening circulation scheme; and Lorenzo Matti, a brilliant architect who eventually led the design development, and I became gradually immersed in the task of physically shaping those incomprehensible diagrams. In hindsight, it strikes me that airport security was never at the forefront of our discussions. Within the context of designing this airport in the late 1990s, the debates on the landside–airside frontier ranged between making the long walking distances pleasurable and showing our understanding of the growing demands of customs and migrating-passenger processing, but rarely on security issues. But this was about to change under the light of the rise of terror in the grim, forthcoming decade.

Only a few years after 9/11, I had the chance to design the new international Terminal B in Monterrey, Mexico. My role as the architect-in-charge gave me a strong position during the formulation, negotiation, and approval of the project layouts. In this case, the general context could not be more different than Toronto; we were forced to comply with the new international regulations—which were harsh with regard to security and the Mexican Federal Police force's codes and criteria.

The atmosphere was usually impregnated with security concerns and the insistent need to create new revenue sources. But it is perhaps thanks to this technical stage of uncertainty that the introduction of changes and innovative solutions was possible. In short, the boundary in Monterrey is quite different than in most contemporary airports today. Instead of following the conventional breaks—laying out horizontally public areas at the landside, *fish tanks* in between, and aircraft at the platform—and vertically overlaying departures (above) and arrivals (below), we opted to introduce a new area which we called the limbo. This new zone was under the main nave of the building, producing a rather dramatic and monumental space between the concourse (public areas) and the holding rooms (fish tanks). The purposes were simple: First, I wanted to balance the passengers' experience at the arrivals (because the idea of airports being the gates of the city was a mere architectural euphemism, there is nothing worse than the chain of low-ceilinged, crowded rooms that await incoming passengers in almost any airport), so we broke that rule, and we opened up the baggage claim hall, giving it the same spatial hierarchy as the check-in hall. Second, we followed the same strategy with the always controversial security checkpoint.

Instead of sticking it in a transitional low-ceilinged space, we pushed it out and into the limbo and displayed it. Just in making these two moves, we challenged too many norms and raised too many concerns. But in the end, the result is fully comforting; passengers appreciate the new stress-less, relaxed atmosphere below the cathedral-like roof. The third innovation that we fought for is a vertical division between the retail zone and the hold rooms. Typically in contemporary airports, shopping is laid at the same level and contiguous to the waiting areas, creating a noisy, hectic, and congested environment for users. Thus, we decided to split them vertically: Retail and restrooms remained at a lower, open level, and the hold rooms and lounges were displayed above, next to the glass wall offering expansive views of the runway. We faced very strong resistance from everyone involved, but we managed to demonstrate that our scheme would work well in terms of retail standards. In all three cases, we broke the mold because we had a hidden design agenda: We wanted to show how we could bring higher levels of comfort to passengers through intelligent modifications that could be introduced at the landside–airside boundary without disrupting the governing mentalities of the airport.

My third experience in designing new concepts for the boundary is through a new plan that I have named "Modernization of the Landside–Airside Frontier" at the alternate airport of Mexico City, better known today as Toluca International Airport, and perhaps soon as the new Metropolitan Airport. Again, the context is rather different from the aforementioned two. The new paradigm seems to be airport revenue. Many airports in many countries are experiencing indeed the rebounds of an economic slowdown or the same crisis that led giants such as American Airlines and Southwest to file bankruptcy.

Although this generalized concern has been driving most of the decision processes in the meetings, our team has managed to offset a new building at the edge of the existing terminals. Because we are pushing back the apron geometry and plane positions, this building had to be necessarily long, narrow, and vertical; however, it also gave us valuable flexibility to reconfigure both the flows and organization of the frontier, without having to deal directly with the previous maze-like one-floor scheme. At the same time, J. Garcia-Bejos, the newly unpacked airport director (who was borrowed from the private transportation sector) proved, besides bearing the commercial pressures, to be genuinely interested in adding to the passengers' experience. This is relevant because when actors negotiate changes at the landside–airside boundary, it is only through the imposition of a clear vision that those changes may be agreed upon by committees and boards and in technical meetings. Like the previous two examples, the new boundary at Toluca becomes a snapshot of its historical context. Thus, the cultural mirror does not refer to expressions of chauvinism or *themed* nationalism; instead, what I argue is that I have found in practice how in the ways landside–airside boundaries are thought of, altered, or modernized, they attend to unique and particular social and cultural demands.

In sum, I have argued so far that landside–airside boundaries—whatever the planning story behind them—are custom-designed socio-technical artifacts that act as cultural mirrors. Returning to the question of why all landside–airside boundaries are different, I have suggested that the major contextual difference (physical, political, managerial, economical) between planning new airports and expanding existing terminals is the degree of freedom in the hands of designers, architects, and planners to come up with solutions that respond to the specific needs of an airport master plan. In the case of expansions, designers must adjust existing terminal buildings and aprons, which are often seriously constrained in terms of room for modifications and additions, and are typically congested. New airports represent a challenging degree of freedom in the hands of planners, although historically, only on three occasions has this opportunity been used to reinvent the airport system from scratch.

In addition, I argue that there are two basic kinds of airport boundaries: domestic frontiers and international borderlines. Although many regional and international airports look the same at first sight, they are notably different, especially in operative terms. Even though all airports must be designed following international norms, regional airports face the opposite problem of congestion: discontinuous flows. These issues complicate the basic functioning of security, airlines, and retail because there are long, dead periods of inactivity. Many local and regional airports remained in single, one-story buildings that could not be connected directly to the planes. In addition, these facilities usually service narrow-body aircraft of smaller dimensions, so passengers are planed and deplaned directly at the apron. Thus, this operative feature dramatically changes the user's relationship with the landside and the airside. In contrast to the obsessive containment policies of large-scale airports—where passengers are treated as *liquids* that should not *spill out*—smaller, regional airports do allow open transit between the two realms. Even though airline employees are usually watching, passengers may feel how the boundary fades off, and the experience of aviation becomes sensorial again. But in regard to scale, huge international airports that choreograph their own internal security policies, few can be so distinctive than Istanbul's Ataturk. For example, a passenger traveling to New York must go through at least five screening filters before boarding and in addition should stay contained in a number of designated areas for quite a number of hours (this may be only topped by adding the mandatory car inspections at Ben Gurion's landside in Tel Aviv). Regardless the nationality, first they should pass all baggage through an X-ray screening, which is right at the doorway of the building. Second, they have to answer a face-to-face interrogatory at the check-in point. Third is by passing through the strict security checkpoint X-rays and ion scanners; fourth passes through a new interrogatory at the gate, one more pass of all handbags through X-rays and one last random control at the air bridge while boarding. In this last stage, passengers will be 'contained' in waiting rooms that are barricaded with planters, screen walls, benches, advertising displays, and everything else you may imagine. If someone needs to use the restroom, they must go through the last steps before coming back to their seats. The paradox of national styles is that Ataturk couldn't free itself from terrorism, and in June of 2016, three suicide bombers created a deadly explosion at the arrivals hall. Why? Because despite the local intelligence displayed at departures, the terrorist group attacked by detecting the relatively weaker point at arrivals. In a city where Isis cells have been identified, the way to think and layout the landside–airside boundary can be way more challenging.

Fig. 2 Rehearsing a traditional serenade, typical in Mexican welcomes, even after short trips. Scherschel, Frank (1958) *Central airport where musicians entertain travelers* [digital] (Reproduced with permission of The Life Picture Collection/via Getty images)

Third and last, boundaries have become a representation of cultural expressions. The permutations among the above-mentioned avert the existence of identical repetitions or transferrable models; therefore, each boundary will capture within its materiality the values, conflicts, emotions, or concerns of its time. However, because of boundaries' constantly changing nature, this representation will always be only a snapshot of a very short window in time. There is though, in the world of airports, a fascinating trend toward *localizing or indigenizing* many airport features (Fig. 2). This is perhaps an escape from the boring, highly impersonal and repetitive place-less-ness so well captured and mocked in Jason Reitman's indie film *Up in the Air* (2009) and sharply theorized by the French anthropologist Mark Augé. One may find in contemporary terminals an assortment of local products, regional information, and even the display of local art. In the end, the landside–airside boundary captures a mosaic of cultures, represented by the sale of Maybach cars at Dubai's airport, the patch of rainforest in Kuala Lumpur International, the smokey taco-stands at Mexico City's airport, the generalized sale of household appliances—such as refrigerators and washer-dryers in Cairo's airport—the fountain-gardens at Changi's airport in Singapore, the hallucinating repetition of statecontrolled products in all shops (including duty-frees) in Jose Martí airport in Havana—excluding the famous Cuban cigars, the massive foot-massage offer at Pudong Airport in Shanghai, or the hardwood, Viking-ship-like finishes at Reykjavik's airport in Iceland. But besides the paraphernalia, cultural expressions are imprinted in complex practices such as the way the USA imposes a national policy to an independent country. In 1952, Customs and Border Patrol (USCBP) established a program particularly intended to avoid congestion at ports of entry in American airports. It consisted in segregating US citizens in foreign territory (airport hold rooms) once they passed security and immigration controls. These voyagers will virtually ended up trapped in a sort of limbo between the local and the American jurisdiction. This process is called international preclearance and since then it has been implemented in Dublin, Shannon, Abu Dhabi, and more widely in the Caribbean and Canada. The major implication of the program is the strict demand of having active US armed personnel in a foreign country. This example shows how far a country is able to stretch the boundary when speaking of national interests or ideology. On a softer side, cultural expressions such as art and history may be found right at the division between the airside and the landside. Just to mention one more example, imagine an art gallery of monumental proportions, let us say a grand museum foyer in the middle of an airport terminal and you will be picturing Vancouver International Terminal (see Fig. 3). The amazing collection on display intends to 'immerse' the visitor into Canadian culture almost before they set foot on Canadian jurisdiction.

Fig. 3 Are airports turning into the new museums? Reinhard, Tiburzy (2017). *The sculpture "The Story of Fog Woman and Raven" by artist Dempsey Bob is permanently exhibited in the terminal of the Vancouver International Airport* [digital] (Reproduced with permission of Shutterstock)

The extraordinary artwork pleasantly distracts passengers from the more standard 'ominous' boundary (martial in atmosphere), making the experience richer and more humane; you may find here original pieces and reproductions of boats, sculptures, totems, artwork, textiles, and even an aquarium! In spite of the unavoidable trans-nationalization of airports, local culture continues to add brushstrokes, mainly in the airport's no-man's land: the landside–airside boundary.

References

Interviews

Dunlay, W. J. (2008, March 8). Transcript. *Personal interview with Bill Dunlay with Leigh Fisher & Associates.*

Publications

Aéroports de Paris. (2004). *Roissy—Charles de Gaulle: 30 ans de déceloppement économique et de partenariat local*. Paris, France: Author.

Aéroports de Paris. (2009). *Terminal 1 at Paris—Charles de Gaulle airport celebrates 35 years and adopts a new image*. Paris, France: Author.

Block, J. V. (1968, December 2). *International airport planning as influenced by aircraft development*. London: Lecture to the Royal Aeronautical Society.

Block, J. V. (1971). *Airports and environment*. Edition Aéroports de Paris.

Brancker, J. W. S. (1977). *IATA and what it does*. Leiden, Netherlands: A. W. Sijthoff.

Froesch, C., & Prokosch, W. (1946). *Airport planning*. New York, NY: Wiley.

Goodman, W. C. (2000). Transportation by air: Job growth moderates from stellar rates. *Aviation Employment*.

Hughes, T. P. (1987). *The evolution of large technological systems: The social construction of technological systems*. Cambridge, MA: MIT Press.

L' ECLAT rédaction. (2011, February 13). *Charles de Gaulle Terminal 1: l'aérogare mal-aimée*. L'ECLAT Regards du Actualité.

Levine, M. E. (2008). *Airport congestion: When theory meets reality*. NELLCO Legal Scholarship Repository, New York University.

Onnée, J.-F. (2004, December). *Paris—Charles de Gaulle Airport: A critique of the development of France's first airport*. Term paper for MIT Airport Systems Planning, Design and Management.

Poh, E. [Assistant Director of ICAO Affairs]. (2007, April 9–13). *Lecture on Strategic Airport Management Programme, Airport Planning and Terminal Design*.

Rogers, E. (1995). *Diffusion of Innovations*. New York, NY: The Free Press.

Salter, M. B. (2008a). The global airport: Managing space, speed, and security. In M. B. Salter (Ed.), *Politics at the airport*. Minneapolis, MN and London: University of Minnesota Press.

Szurovy, G. (2003). *The American airport*. St. Paul, MN: MBI.

Washington National Airport. (1941, October). *Architectural Record, 90*(4).

6

"The Liquid Airport": Security, Permeability, and Containment in Airports

> "the door of arrivals" and "the door of departures" … "Between the two the meantime is all there is. It passes of itself."
>
> —Allen Curnow (Poet), "Narita," in *Early Days Yet: New and Collected Poems 1941–1997*, p. 32

This is the way Allen Curnow (1997) describes with great subtlety, the fact that the airport is also a game of controlled entries and exits that will inevitably lead into the "in between" or what can also be represented as a series of spaces that have been "paused" from the exterior's rules and realities. In this sense, as users increase flowing globally, airports resemble liquid's containers, which, depending on location and culture, may feature certain degrees of *permeability*. Metaphorically speaking, this fifth chapter discusses the growing relevance of the landside–airside boundary as a fine-grained filter that controls the density of airport terminals, and, eventually, the flows of those liquids that try to pass through. If it is true that passengers and employees need to flow through airports like liquids—in addition to the infrastructure's natural *porosity*—designers who were increasingly pushed to look for separation and new strategies of containment started envisioning the airport as a system of control *valves* (Fig. 1).

V. Marquez, *Landside | Airside*, https://doi.org/10.1007/978-981-13-3362-0_6

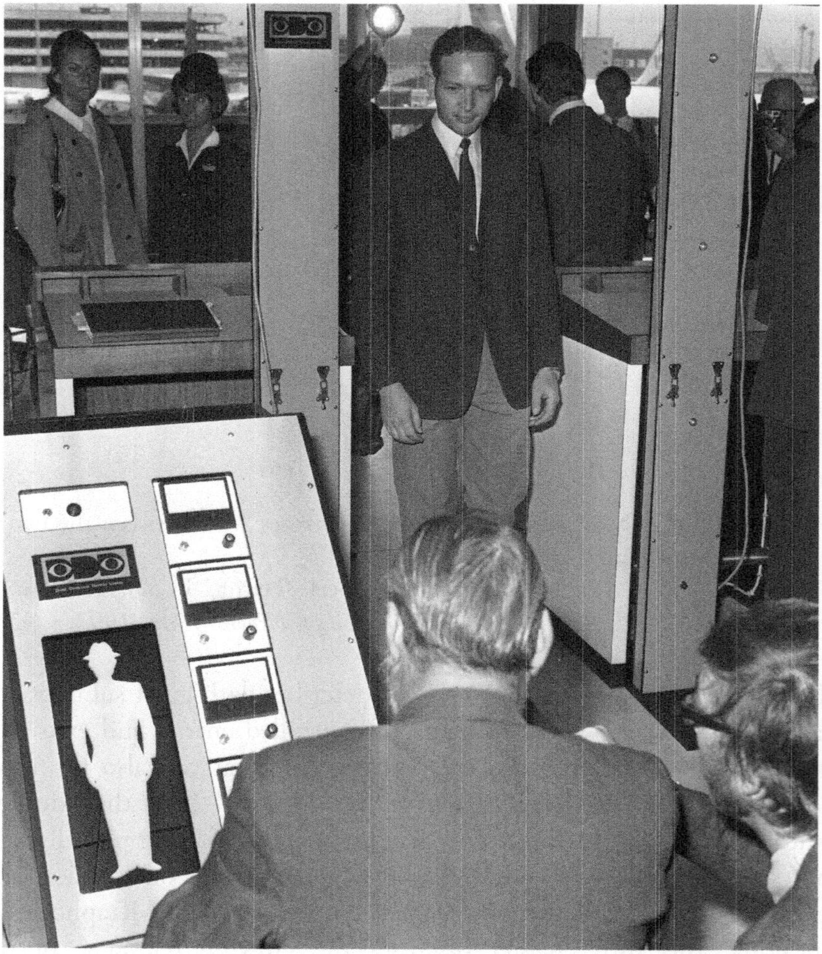

Fig. 1 Cutting edge technologies, like metal detectors, are used for the first time in London's main airport. Observe the gendered connotations of the device. *New Anti-hijacking* [digital] (1970) (Reproduced with permission of Popperphoto/Getty Images)

As I present in the next sections, there is evidence that the larger discourse of society, particularly upon the rise of international security concerns in the mid-1970s, has had a tremendous effect on the way the landside–airside boundary has more openly become the main feature of airports. As internal and external need for security grows, so does the need to filter, sort, and split more efficiently and perhaps even more drastically. Paradoxically, the increased need to enforce strict filtering was accompanied by an array of devices and routines, ranging from non-invasive new technologies to the extreme of displaying heavily armed military presence.

As I have mentioned earlier, much of this chapter is framed around the challenging context of planning and designing landside–airside boundaries after 9/11. The huge impact of these events brought a strong critical response, which has been thoroughly documented in a new literary production on airports, much more committed to a social, political, and legal agenda. I argue that most of this new body of literature spins around landside–airside boundary conflicts, and that most authors question the very nature of the airport and its limits, past and future use of technologies, the regulatory frames, and the rising pressures of retail.

In a different sphere, I argue around the fact that the boundary has become a strong mechanism of public coercion and control, including the ways it materialized as physical requirements that had to be attached to—or often forced into, existing terminal buildings. But how permeable are these thousands of airport terminals under the new standards? In this sense, non-material technologies started playing a decisive role in compensating for those buildings' physical limitations; thus the construction of an immaterial frontier was ominously imminent. I maintain that there are at least three new kinds of landside–airside boundaries: the psychological constructs, the informatics frontier, and the establishment of virtual mechanisms of control. In short, this chapter offers reflections on the benefits of making the invisible visible. So, to make history useful, what could the role of landside–airside boundaries be in thinking of airports of the future?

1 Discursive Aspects: The Paradigm of Security, Rethinking What Airports Are

The alarming rise of hijackings of the late 1960s reached a peak of 89 cases in 1969. But until then, most of the piracy had political purposes and was meant mainly to draw the attention of the press in order to gain publicity. Moreover, in 1972, blood was spilled for the first time in a massacre that led to the death of 24 people at Israel's Ben Gurion International Airport. It was perhaps when Raphael Ron, BGIA's head of security at the time, declared "that terrorist attacks are carried out by people who can be found and stopped by an effective security methodology" (Khurana, 2009, p. 271) that airport security became the new paradigm at the airport. Since then, the boundary has been re-elaborated endless times within airport planning. The added complexities of security became such an exponential factor that throughout the following three decades, they would build a labyrinthine and massive apparatus of international jurisdictions, new technologies, protocols and procedures, identification standards, specialized agencies, and so forth. Paradoxically, the new security system was nested right at the landside–airside boundary. Thus, during the following years, just knowing what the boundary was, or even figuring out where it was, became increasingly a detective's task.

In the relatively encased history of airports, one major externality would once again reshuffle the relationship between airports, aircraft, and their users: the events of 9/11. It was right at the landside–airside secure frontier of the airport where radical men equipped with a well-orchestrated plan showed the weak side of the system and turned civilian aircraft into military weapons. Thousands of innocent lives were lost. As a consequence, the airport immediately was put in the spotlight, thoroughly scrutinized, and re-regulated. Since then, romanticism has been replaced by discipline, and the airport has now grown into a living fortress. As time moves on, the haughty travelers of the past have faded away. Today at the airport we see only long lines of "docile bodies" (Foucault, 1977, p. 135) ready to queue and undress, separate from their belongings, follow martial instructions, and observe something that resembles a rhetoric of fear (see Fig. 2). Sometimes the larger discourse of society reaches out to the farthest corners of technological change, but without a doubt, few paradigms have been so extensively influential in aviation, airports, and, finally, in shaping landside–airside boundaries. In the end, a "cost-benefit analysis of the interests of multiple stakeholders engaged in aviation security offers some empirical

Fig. 2 The dance of the "docile bodies." Dobilas, Marius [digital]. *Security and passport control at Antalya Airport* (2017) (Reproduced with permission of Popperphoto/Getty Images)

insight into the stakes involved and the logic behind the border control" (Lahay, 2008, p. 78).

Pushed by the USA, most countries realized that airports were the new national borderlines, and that those border points may put their own national security at stake. The new criteria imposed by *Annex 17* (ICAO, 1974) made its way to every single airport within the international network. Even small local airports were forced to follow the new tight regulations. What really cornered airports was the fact that their buildings, or infrastructure, not only were not ready to allocate the new requirements, but, furthermore, were sometimes playing physically against the implementation of those new rules. This condition unfolded the landside–airside boundary as a new physical and virtual barrier. But if there was ever any clarity in identifying landside–airside boundaries in airports after the terrorist attacks in New York and the emergency measures taken by the FAA, ICAO, and later the international community, the border became simply fuzzier than ever. Thus, planning and designing airports and terminals became much harder. Even the adaptations at facilities that were built before the turn of the century made little sense on paper and far less sense in reality. Simply, most terminals around the world were not equipped to allocate the new improvised configurations of the boundary; hence, users' discomfort again grew exponentially.

But after terrorists used aircraft as military weapons, the security systems of airports were spotlighted and questioned. In the aftermath of the tragedy, many specialists started questioning the very nature of the airport. Writers, lawyers, sociologists, psychologists, intellectuals, and others who engaged in this topic, particularly in the aftermath of the tragedy, initiated a polemic around the meaning of airports, and started offering fresh visions of what an airport could be.

As I discussed in earlier sections of this work, back in the days of New York LaGuardia or Paris Le Bourget, "airports thrived on being social and local rendezvous for plane spotting and air shows" (Gordon, 2004, p. 55). Airports were equipped with "observation terraces open to the public, the piers along which the great jets docked, along which, every Sunday, families would range themselves with picnics to spend the afternoon contemplating the movements of aircrafts" (Pascoe, 2001, p. 55). During the decades after 1930, "modernist architects influenced by Le Corbusier, wanted to create feelings of drama—an increased connection with the theatricality of flight would see airport buildings disappear into an overexposed space of glass" (Pascoe, 2001, p. 35).

The best example would be the terminal at Orly in Paris,[1] which incidentally may be considered the opposite of TWA. For David Pascoe, a professor of literature and an author on the sociology of the airport, it was not until Reyner Banham's "influential and prophetic article The Obsolete Airport" (1962, as cited in Pascoe, 2001, p. 134) that planners and administrators realized that the changing nature of the airport was a complication for the future. For example, the broad adoption of

[1]In late February 1961, DeGaulle opened the new terminal at Orly and this example would become key to understanding the planning and design philosophy of the new European era. "Orly's architect, Henry Vicariot, created a terminal consisting of a massive rectangular box frame fabricated out of stainless steel over which, in order to absorb ambient sound, curtain walls were

wide-body and stretch jets during the 1970s (between 100 and 300 passengers), drastically increased the number of passengers and proportionally the size of airport terminals.

As Mark Gottdiener (2001), a sociologist interested in airports, argues, "their introduction (twenty) years ago necessitated the alteration of nearly every airport in the country" (pp. 64–65). Within this same line of changes, Anne Graham (2001), an expert on airport management in the UK, argues that during the 1980s "the airport industry was not immune to the *quality revolution* which was taking place—although it was rather late in adopting some of the principles. Structural changes such as commercialization, privatization and globalization, together with increased competition between airports, encouraged airports to place more emphasis on quality" (p. 72).

Moreover, several added complexities to this constant of change shook the airport at its roots. For example, Mark Salter (2008), a researcher on political studies and an influential analyst of airport borders and identity processing, thinks that "new aviation security requirements and privatization have dramatically altered the management of global airports" (p. 22). This is perhaps true for every single facility in the world, and along this same vein, it has also altered the multiple ramifications that lay within. Thus, according to Salter, there may be "three major dilemmas within contemporary airport security: space, speed and sharing" (p. 12). In terms of space and speed, airports perform as "facilitators of movement; they act as tunnels or passageways that create clear and free-flowing routes to the aircraft" (De Neufville, 1975, p. 23). Although the "airport terminal itself is striated by various impediments designed to shape and control movement[, t]his is reflective of the wider airport functionality of a machine for capturing and controlling flows"

elastically suspended. For the first time since the war, Europe seemed, perhaps, to be capable of creating its own image of modernity, drawing on its native canons and interacting with the present on a more objective basis than, say, Saarinen's TWA terminal at JFK. For Orly was, in effect, a massive *vitrine*: a window on the world, and a display case. This extraordinary terminal contained a new world consisting of a 300-seat cinema; several hotels; an exhibition space; a Michelin-starred restaurant; a chapel fitted out with marble detailing; a crèche strangely decorated with brightly coloured birds; and, of course, shops stocking luxury products from all over France and its colonies" (Pascoe, 2001, p. 54).

(Fuller & Harley, 2004, p. 14). As buildings started to be designed or redesigned after 9/11, the "metaphor of the filter achieved material form in the shape of the airport terminal itself" (Adey, 2008, p. 156), though perhaps becoming gradually more disconnected than ever from the rail station and other predecessors.

For some analysts, the new "terminal is organized to move passengers through to a boarding gate or lounge with simple and direct flows with little conflict or interruption" (Doganis, 2004, p. 139). The dual pressure of congestion and security created a paradox. If planners *channel* people's movement and redirect them where authorities want them to go, necessarily many aspects will enter in conflict: for example, higher enplaning and deplaning delay, longer queuing times due to detailed scrutiny or interrogatories (as in when flying into Ben Gurion in Israel), passengers forced to follow consumption paths, and so forth.

In sacrificing monumentality and passenger comfort, many airports have turned into "systems that are dense, fast, and contingent" (Salter, 2008, p. xiii). This density is "produced as a result of the combined activities of various different organizations such as airlines, handling agents, customs and immigration officials, concessionaires and so on" (Graham, 2001, p. 72). Managing security seems to involve a growing number of issues, ranging from political internecine wars, social debates on civil rights, debates on national security, or the exponential economic and financial complexities of keeping healthy balances without receiving subsidies. I argue that this fragmented complexity makes it simply much harder to read the airport's new processing demands.

In addition, "the governance environment for any particular global airport is a combination of international treaties, national regulations and legislation, local bylaws, and management practices" (Lahav, 2008, p. 16). Thus, it is evident that airports work as links within a larger chain of actors and interests, reinforcing Deborah Douglas's (1995) idea of airports as systems embedded within larger systems, or institutions dependent on global, socio-technical networks. Therefore, this threaded interconnectivity implies that the airport security system will always be tied to the least secure airport of the network, thus ensuring that under this condition airports can hardly be standardized.

As I have argued before, there is a long and uneven relationship between airport users and multiple technologies—not just airplanes, which are centralized in terminals. As George Scullin, the author of the "International Airport" (1968) maintains, "in bringing together this ultimate collaboration between men and men, men and machines, one other source of power has always been in the forefront" (p. 23). But as the airport is subordinated to larger political discourses, technologies appear as suitable mechanisms of social scrutiny and coercion. As Fuller and Harley (2004) suggest, "the question of path finding has become more pressing as terminals grow in size and complexity" (p. 127).

This is especially true for airport designers. But who is really behind these initiatives? Are planners, government officials, and regulator bodies, perhaps? Salter (2005) has defined the airport as a "transient institution" that has "three spheres: the arrivals hall, where newly landed passengers sort themselves into citizens, foreigners and refugees; the 'confessional,' where the border agent inquires about your provenance and destination; and the passport check (also by the border agent). This last sphere is the 'hyper-documentation' phase where the 'body-biometrics-file-profile' is central" (Salter, 2005, p. 45). But I would like to point out that Salter leaves aside perhaps the most significant "sphere" of all: the *limbo*, or the zone of no transition. Salter perhaps overlooks that in fact passengers may change status basically everywhere in the airport, except within the limbo. Thus, I maintain that although the *processing* of passengers is undeniably related and dependent on a complex legal, technological, and jurisdictional network, it is the controlled zone, or what I call the limbo, on which the airport's future depends. Anne Graham (2001), a researcher of tourism and transport at University of Westminster, speculates:

> … at most major hubs there will also be passengers who spend a considerable length of time in the airside area.
>
> Various airports have developed some quite imaginative airside facilities and services which can be shared between these and local passengers. For example, Singapore's Changi airport has a swimming pool, a sauna, a karaoke lounge and a putting green and, if the passengers stay longer than four hours, they can go on a bus tour of Singapore. Amsterdam airport

> has an art gallery and casino. Vienna airport opened its Wellness Plaza in 1999 where passengers can experience different massage techniques and make use of the fitness bar, solarium and showers. Bangkok airport offers head and foot massage, Frankfurt airport provides dental, optical, and foot reflexology services and a clinic which has become a specialist centre for laser operations to repair damaged shoulder and knee joints! (p. 129)

I tend to disagree with Graham's (2001) use of "airside," because I believe she refers to the controlled zones, which mediate between the landside and the airside, and therefore are described in this work as the boundary, or the limbo. Nevertheless, and in spite of what I dispute, it is fascinating to see how the boundary becomes step-by-step more *elastic* through the years. Are we witnessing perhaps the birth of a new, Foucauldian kind of architectural genre? But if on the one hand the new security paradigm forced the physical containment of landside–airside boundaries, on the other hand it has pushed technical innovations in two directions: First, the construction of "certainty," or the development of precise and reliable equipment that could take out of human hands as much responsibility as possible, and second, the construction of surveillance systems, or the implementation of sophisticated devices for institutional vigilance. Hence, boundaries will carry on the added complexities of implementing sophisticated technologies in order to achieve the "public expectation for absolute security" (Salter, 2008, p. 1). And this is precisely why the meta-discourse of national security has relied on artifacts. The public perception of a safe terminal seems to be paramount in the post-9/11 environment; as the burden of guilt has been placed on the incapacity of security officers to identify terrorists at Logan Airport, institutions have bet on the public reliability on automated technologies.

Even though the landside–airside boundary can be "drawn in terms of information" (Bennett, 2008, p. 4), it does not mean that the negotiation network could remain closed. In this sense, Gallya Lahav (2008), a researcher on political science at SUNY Stony Brook, argues that "the introduction of innovative technologies, biometric techniques, and new visa and passport procedures has extended the number of regulatory actors beyond national borders" (p. 88). This is especially complex when technologies are operated under different levels of skill and cultural background.

Finally, the last technological outcome after 9/11 points toward the matching of databanks and physical characteristics of people. It has been documented that using "Disney's expertise with biometrics, [...] at least one of its executives advised the Federal Aviation Administration on airport security after 9/11" (Harmel & Spadanuta, 2006, p. T9). But this new technological strategy will unavoidably imply multiple questionings. Not surprisingly, landside–airside boundaries have become, since then, physically responsible for facilitating passenger sorting. In this sense, "contemporary concerns about the privacy of the physical person and its protection from various biometric devices are also centered on a notion of physical or spatial boundary" (Bennett, 2008, p. 3). Thus, a more complex interrelationship between overlapping and parallel borders is foreseeable, although these new technologies are planned to be as removed as possible from users' presence. Will high-tech scrutiny be contestable in the next few years?

2 Normative Aspects: Coding a *New Air World Order*

In the aftermath of the terrorist attacks, deep institutional and regulatory reforms could not wait. For example, International Civil Aviation Organization (ICAO) issued both temporary and permanent amendments to its codes and regulations, and the USA reacted with the creation of a super-sized ministry, the DHS (Department of Homeland Security)—which came to be mainly in charge of screening and processing passengers at airports and other ports of entry to US territory. But perhaps the most sound of all measures was Amendment 10 to *Annex 17*, which was adopted by the ICAO Council on December 7, 2001, in order to address challenges posed to civil aviation by the events of September 11, 2001. It became applicable on July 1, 2002. The amendment included various definitions and new provisions in relation to the applicability of this Annex to domestic operations; international cooperation relating to threat information; national quality control;

access control; measures related to passengers and their cabin and hold baggage; in-flight security personnel and protection of the cockpit; code-sharing/collaborative arrangements; human factors; and management of response to acts of unlawful interference. Among the myriad of new programmatic requirements, regulations, and recommendations, several new terminals grew immoderately. In this regard, "airlines have also found it necessary to sound a note of caution about the construction of new terminals which appear to be in excess of requirements because, although palatial buildings may enhance national prestige, the cost must ultimately be reflected in the landing fees and other charges and thus influence the level of fares and rates."[2] In the early 1960s and until the mid-1970s, in most American airports, airlines were paid for their own terminals, especially in airports following the "Airport City" model of Idlewild. Although this practice decreased during the following decades, in the late 1970s, for IATA (International Air Traffic Association) it was the airlines who should establish some "general principles" in airports, although it was then an airport terminal advisory committee composed of airline experts who had effectively set out the important factors to be observed by the authorities concerned with designing and constructing new airports.

It would be impossible to produce a design appropriate for every purpose and locality—each airport serves a different type of traffic and a different mixture of airlines, and must be specified accordingly—but the general principles are common, even if the needs must be met in a variety of ways. In consequence, the lack of private investment drastically slowed the planning of new terminals, and "among major metropolitan areas only Denver had managed to build a new passenger airport since the 1970s" (Altshuler & Luberoff, 2003, p. 279), and most airport projects consist in expansions and remodeling of existent facilities. "Without in any way denigrating the expertise of the airport authorities, it is the airlines who are the essential users of airports and they are the people who face the major problems and ultimately pay the bills" (Brancker, 1977, p. 23). Thus, at the rise of security issues, IATA leaves

[2]IATA has produced an airport terminals manual (Lyon, 2007, p. 109).

clear that airlines will fight for regulations due to the impossibility of standardizing airport design. However, the new codes and additions will respond now, beyond listening to the key players in the industry, to the larger paradigms of societies who are part of the global aviation and airport system.

In this sense, and at least on paper, ICAO was the only organization capable of launching global reforms. Among the major changes, *Annex 14*, Volume I, of the Chicago Treaty, has been subject to thirty-nine amendments since 1947, and "*Annex 17* in full has been revised twelve times since 1944 to adapt to new risks and challenges, shifting focus since the 1970s from hijacking to attacks on facilities, to sabotage, including the use of aircraft as weapons of mass destruction" (Salter, 2008, p. 17). In this regard, ICAO needed to act fast and standardize as much as possible from current practices, but especially the ones related to passenger identifications. "In relation to the 'war on terror,' one of the most obvious signs of global surveillance is the agreement of the ICAO for its 188 contracted states to use standard machine-readable passports" (Lyon, 2008, p. 44). These measures and requirements that frame "the efforts of a capable international body to develop technical standards with the potential to improve air travel safety seem laudable (see Fig. 1). Although legal scholars, such as Osieke" (Stanton, 2008, p. 3), have questioned the ICAO's authority to set and enforce international policy, Salter (2008) has suggested that,

> ….in terms of secured space, airports are divided into public and restricted areas. Sterile areas are those public spaces that have been security screened. Restricted areas cover all nonpublic spaces for which authorization by the National Civil Aviation Authority is required. Landside operations refer to those within the terminal; airside refers to spaces with airplanes such as runways, taxiways, and aprons. However, airport security regulations have changed dramatically in response to terror attacks. (p. 13)

It is interesting that the author implicitly holds that the divide between sterile and non-sterile does not correspond to the landside–airside boundary.

For instance, after the attacks in Rome and Athens in 1973, the FAA imposed new rules: "narrow points of control were established at the 'throat' of each concourse" (Gordon, 2004, p. 13). This is the case with facilities such as Tampa, just to mention an example, in where the terminal was not equipped with pre-filters or filters. Within the new coding order, an endless number of airports had to cope with new demands such as having to deal with no room for expansion, reducing airport income due to recession, and dealing with new regulations, procedures, and new protocols.

3 Constructing an Object of Knowledge

Are there any features that become visible only through the use of a landside–airside boundary as a lens? Before I tackle this question, I propose to structure my response with some methodological order. I will first introduce how the boundary may be used as a research tool, and how certain cultural features consolidate the same boundaries as independent and autonomous pieces of knowledge. Mindful of the fact that much of the evidence product of airport negotiations have remained vaulted for decades, I have sought alternate ways to bring to light what otherwise was forgotten in company and public records. Possibly, the main finding of this book is the identification of the landside–airside boundary as a tool for opening those black boxes. If this is so, it could be interesting to test the same strategy for other cases alike. With this idea in mind, I have looked for a word that may describe the process of identifying entities (generally defined by those mediating the tensions between the interests of users and artifacts) and isolating them, in order to make visible the true reasons of technological change. I propose to borrow from computation the concept of *viewport*, which seem to capture most of the features displayed by landside—airside boundaries—at least for the purposes of this book. So technically a viewport is a bi-dimensional window in a three-dimensional space. They have the property of revealing critical information that makes hidden decision-making visible. Therefore, I propose that the identification of technological viewports could help other researchers in a similar way S&TS

Table 1 The five epistemic features that may shape a technological viewport

Epistemic feature	Description
Discursive	Paradigm of security brings about rethinking what airports are
Normative	New codes establish a new set of rules for drawing landside–airside limits
Object of knowledge	A new literature addressing the landside–airside boundary appears
Transitory	Boundary spaces are characterized as a sort of no-zone or *limbo*
Intangible	Boundaries are significantly virtual and psychological

scholars have succeeded in opening black boxes through the study of crises, accidents, and catastrophes, etc. Also interesting is the fact that technological viewports are not just interfaces (or the physical devices that users employ to interact with machines), but "windows" on the intermediate spaces between humans and technological systems. MIT's Sherry Turkle (1995) has written that in computer culture there are two kinds of expert users: those who inhabit the "obscure world of computer *microworlds*" and those who "enjoyed the sense of nothing standing between them and the 'body' of the machine" (p. 32). In computation, at one end are users, who interact through interfaces and inhabit virtual realities—in-between spaces or software—and at the other end is the machine, or the hardware. In large technological systems, we have created similar spaces of mediation, or *limbo zones*, which remain highly overseen by specialists. I argue that it is only through the use of viewports that these entities materialize and become subject to scrutiny and investigation. In the case of landside–airside boundaries, there are at least five epistemic features that suggest the consolidation of a technological viewport (see Table 1). In the next few pages, I discuss each of them in independent sections.

The concept of the viewport creates an artificial sense of reality, just like the three-dimensional environments described by Turkle (1995). Thus, they become rapidly equipped and furnished with all sort of elements that give the idea of an independent micro-world. Users will enter those spaces and temporarily dwell in the in-between. Just as computer users input a security password at the interface, airport passengers cross through the different thresholds of airport interfaces, such as the check-in counter or the security checkpoint. Once inside, they are in neither the human world nor the world of machines, hardware, or artifacts. These contained spaces are bounded by limits and restrictions that condition the transit of humans and products toward either the landside or the airside.

In the twilight of the 9/11 events, a new social literature of airports gradually came to light. Airport coding after the terrorist attacks rapidly commanded the attention of experts on psychology, international relations, security, and national defense and intelligence, as well as journalists, social analysts, and lawyers. Planners and designers also started discussing more openly the different forms of material limits, and the first publications of this sort started to appear. This phenomenon is particularly important to consolidating the importance of the landside–airside boundary as a historical "viewport."[3] I argue that the public acknowledgment of the frontier formalizes its hierarchical importance as an airport definer. In fact, the lack of specialized literature and public criticism is what perhaps buried the boundary as a central aspect in airport design.

The literature that I discuss in the next few pages comes from very different origins, including academics, experts, journalists, and others. I found at least three aspects at the center of the analysis that may represent the main lines of concern: first, the materiality and immateriality of the airport's internal frontiers; second, the contradictions between the dynamics of transportation flow, and third, the slowness of filtering the national frontiers (not surprisingly, all three directly address the landside–airside border). Now I move on to comment on those voices.

The most immediate reference that we may have as travelers is the physical existence of limits in airport terminals. In general, "the notion of border is very often considered a materialized line between two spaces" (Bigo, 2004, p. 52). In fact, physical borders "create a membrane or buffer zone separating an inside from an outside, while

[3]This concept may also be understood as a variation of *enframing*. In other words, I introduce "viewports" as imaginary, two-dimensional windows that are focusing on hidden technological relationships—usually created during a formative process. When these networks are augmented and unfolded, they usually reveal much of the essence (or true purpose) of these objects.

linking both in a particular way, projecting the imagination of a larger, encompassing reality on the ground" (Passi, 1996, p. 3). But in the case of the airport, the most important boundary is the one that divides national territories. In the book *B/ordering Space* in the *Border Regions* series, authors Van Houtum, Kramsch-Karsh, and Zierhofer (2005) refer to territorial border as a normative idea and even play with its linguistic structure, splitting it as b/order. They also argue that the frontier becomes "a belief in the existence and continuity of a territorially binding and differentiated power that only becomes concrete, objectified and real in our own everyday social practices" (pp. 2–3).

Although it has been said that the "world is indeed becoming 'liquid' or 'borderless'" (Van Houtum et al., 2005, p. 2), perhaps "borders can be imagined as lines of difference and protection as well as representations of the crossing, the de-fencing, the merging, the additional or the expectation" (Van Houtum & van der Velde, 2004, p. 100). Given the apparently contradictory spatiality of borders—which is perhaps best reflected in their capacity to articulate both transcendent closure and immanent openness—we may assert that bordered spatialities are partial, selective, and opportunistic both in their representation and in the interests that they serve.

However, the physical representation of boundaries is often charged with tangible intangible messages from higher powers and institutions. Borders are often indicative, inductive, and sometimes invasive. Since 9/11, concern with small things, such as small knives, has been paramount, even to an obsessive degree (Lyon, 2008, p. 42). Unlike the pre-1980s control points, contemporary borderlines such as the security checkpoint have transformed into primarily material filters.

Although borders and airports are clearly meeting points of social systems "that function as spatial mediators of often latent power and governance discourses" (Van Houtum & Naerssen, 2002, p. 129), those discourses materialize inevitably as objects and systems, which eventually require physical presence. But beyond its materiality, according to Salter (2003), the layout "of the space of the passport control area parallels the legal and psychic journey from being controlled and contained at the border to the comparative freedom of the safe domestic space" (p. 126).

This speaks of a level of transformation between the physical and the non-physical. Although for Salter (2006), "the effect is that the airport is rendered immaterial by the metaphorical abstraction of the filter—constituted by software and informational systems" (p. 147), the materiality of the physical processing of passengers demands equipment and infrastructure. In addition, the filtering and surveillance of things include supplies, baggage, documents, aircraft, and so forth. But key in the understanding of borders is not only their material form but the degrees of representation and interpretation that they embody. As the French sociologist Bruno Latour (1993, p. 3) has said "a key can transform a door into a border for some but a pass-through for others, in the same way that a wall may signify protection for some while for others it constitutes a political offence or merely an irritating graffiti board."[4] Thus the frontier could be further explored in terms of invisible shapes, structures, values, or interpretations.

Among the immaterial entities that shape today's boundaries, we find data, human perceptions of space and information, human behavior, and decision making. David Lyon (Lyon, 2007, p.132) makes an interesting claim, asserting that "the 'border' is not only virtual in the sense that airports are seldom at the physical edge of territory but also in the sense that processing often takes place 'upstream' of the actual location where documents may be checked." For Colin Bennett (2008), a political science expert at the University of Victoria in Canada, as the boundary can be drawn in terms of information, there are at least "three overlapping dimensions of the problem: humanistic, political, and instrumental" (p. 4). Right at the crossroads of this multidimensionality, I suggest that users have also placed their *intimacy*. The landside–airside boundary is now featured by learned choreographies from passengers, in where users must yield to invasive security procedures that involve—among many other things—tactile, physical searches. For example, Bennett (2008) also argues that these "intimate issues can never be wholly private. Neither can the control of personal information. Whether drawn in spatial, behavioral, decisional, or informational terms, each of these

[4] Latour's idea is intermingled with Ley and Cybriwsky's. The first part belongs to the French sociologist and the second to geographers (Ley & Cybriwsky, 1974, p. 493).

boundaries is inherently flexible, contestable, and dependent on context" (p. 4). The issues on loss of intimacy, invasiveness, and passengers' rights speak of the contradictory nature of the post-9/11 paradigm.

Users are requested now to surrender their own privacy rights in terms of personal information and in order to give feedback to a larger, more Orwellian system of national/international security. This relationship is now conducted through a process that consists of verification (machines and devices) and verification plus interpretation (officials). After 9/11, the processing of embedded data in ID documents became the first line of national defense. One of the most influential new voices is David Lyon, a sociologist who is also the director of the Surveillance Studies Center at Queens University in Canada. Lyon is better known for his deep studies of global mobilization and the strong challenges that this phenomenon has posed to personal identity. For the sociologist and lawyer, airport boundaries are becoming less physical. In this sense, Lyon has claimed that "one's identity as a citizen is checked by these means and others, such as national or category-specific ID cards, which also bring the 'border' to other locations than the 'edges' or territory, thus further virtualizing it."[5] Hence, the dematerialization of boundaries creates an effect that places users not just at the edge of a nation, but at the edge of its integrity. Landside–airside boundaries have gained then, perhaps in different degrees, a military undertone (Fig. 3).

Mark Salter, whom I have cited several times before in this book, is a professor of political studies at the University of Ottawa and perhaps one of the foremost visible scholars interested in building this new literature of airports. His book *Politics at the Airport* (2008) is becoming widely discussed in airport-related environments. For Salter, airports are organizations of constant activity and movement—wherein people's mobility is always tied to "fixed institutions, such as territory borders, customs and border posts, traffic laws, infrastructure, and passports, the state also described 'normal' routes of movement and acceptable modes

[5] Lyon (2008, p. 32). Travelers of all kinds are emblematic of these multiple mobilities, and their data are embedded in passports, visas, and other identifying documents; are read by machines and officials; and are passed from place to place the world over. One's identity as a citizen is checked by these means and others, such as national or category-specific ID cards, which also bring the "border" to locations other than the "edges" or territory, thus further virtualizing it.

Fig. 3 The civil airport is now turned into a militarized facility. Frame Stock Photographs [digital]. *Airport security apprehending a suspect* (2015) (Reproduced with permission of Shuttersock Inc.)

of fixity" (p. xii). Both the internal and external pressures at the airport were reflected in "aspects of security, scarcity, profitability, accountability, and governance [that] create a dense, overdeterminated, chaotic, networked topology where dominant impetus is the management of speed" (Salter, 2008, p. 9). But in this sense, facilitating speed after 9/11 is perhaps more based on the "segregated efficiencies of transportation planning in general in which the grail is the separation of means and, thereby, of people and privileges," as Lyon (2008, p. 40) has described.

In reality, only a few privileged travelers flow faster through airports. The film *Up in the Air* (Reitman, 2009) poignantly portrays the rise of a new social class, which for one reason or another forms part of an elite of speeding up airport processes. As shown in the film, surprisingly, this social group is somehow cohesive and does share an array of international practices for speeding up airport processing. But for the common traveler, living "in a fast-paced, mobile world, the idea of borders as barriers to movement is an unacceptable irritation, but in a fearful and unsafe world such borders make a lot of sense" (Lyon, 2008, p. 42). On the same line of thinking, Christopher Schaberg makes a comprehensive and somewhat parodic account of the emotional pilgrimage of transiting through airports and its cultural implications. The author describes how "*airportness*, or how the feel of air travel precedes and extends past the more obvious dimensions and boundaries of flights" (Schaberg, 2017, p. 3). Is the airport really turning into a social syndrome?

In the post-9/11 world speeding up passenger processing is beyond necessity, a practical fallacy. The rising need to defend national frontiers passed from the territorial border to the room of customs. As Benedict Anderson (1983) portrays in *Imagined Communities*, a "nation is imagined as limited because even the largest of them, encompassing perhaps a billion living human beings, has finite, if elastic, boundaries, beyond which lie other nations" (p. 50). A paradox in practice, though, is that "spatially, one enters the border zone of the destination airport before leaving the departing airport," as Salter (2008, p. 9) clearly describes. For Anderson, the "delocalization of international borders forces air carriers and foreign airports to be the front line of border control" (quoted in Salter, 2008, p. 9), which is true and applies directly in the way contemporary landside–airsides must bring in all the complexities of rethinking global frontiers. Nonetheless, "borders help us to understand the imprecise fit between nations and states" (Wilson & Donnan, 1998, p. 25). Finally, Salter (2007) argues that before the international laws, airports "are national spaces that connect to international spaces, frontiers are not at the territorial limit, and grounded sites that embody mobility" (p. ix). The question is, are we truly leaving national grounds when we pass emigration?[6]

Although this and many other questions may remain open, the open discussion that I have commented on this section truly benefits not only specialists but even a larger audience that is gradually more interested in fly-rights, consumer protection aspects, and airport usage in general. Not until the emergence of a post-9/11 literature on airports reached the far corners of sociology, international relationships, law, psychology, economics, and other sciences related to the humanities has the landside–airside boundary become situated in the spotlight of the debate.

[6]"Emigration" is a standard procedure in Europe and other countries but not in the U.S.

4 Transitory Aspects: The Limbo

Though, beyond the engagement of a social critique, few stories capture best the notion of *placelessness* and conflicts between the landside and the airside border than the case of *Sir Alfred*, or Mehran Nasseri (or the person who lived illegally for 18 years in Charles de Gaulle's *in between* zone due to an immigration problem and whose story inspired Steven Spielberg's film *The Terminal*). Speaking of *Sir Alfred* (see Fig. 4). Justine Lloyd (2002) from UTS in Sydney describes in *Borderlands*, a specialized journal, how "the legally-defined extraterritorial zones—in which the figure of the stateless person has materialised—are a sign of the need for global trade centres and airport cities. These ambiguous extraterritorial zones produce equally ambiguous extraterritorial subjectivities" (p. 1). The Australian author shows "how efforts to contain such extremes of transnationality within border zones have

Fig. 4 Limbo is best represented by reality, how someone end up living nowhere? Fougere, Eric [digital]. *Sir Alfred Mehran* (2004) (Reproduced with permission of Corbis Entertainment/via Getty Images)

produces additional sites of statelessness and un-sovereignty" (p. 1). In her chronicle, Lloyd asserts the relevance of the boundary, even in the unfeasible or perhaps unimaginable case of a passenger living for more than a decade in what I define here as the *limbo*. With no doubts, the best representation of this *no-man's-land* and its inherited vagueness is the fact that this Iranian citizen survived for eleven years in Charles de Gaulle airport, between an electronics shop and a pizzeria.

The limbo becomes a representation of what Marc Augé (1995)—the famous anthropologist—calls the ultimate "supermodernity." For the French author, "If a place can be defined as relational, historical and concerned with identity, then a space which cannot be defined as relational, or historical, or concerned with identity will be a nonplace" (pp. 77–78). In fact, the limbo has produced its own segregated culture within its own little micro-cosmos, for example, cultural constructions such as airport food or airport literature. Bianca Leggett (2012) has claimed that "airport novels are designed to distract the reader from the boredom and discomfort of the airport itself, they provide an escape hatch into inner-space through sensational stories which help to combat the nullity of the non-place in which they are read. While traditionally the 'airport novel' is a place of refuge, the antithesis of the airport itself..." (p. 1).

But beyond the public perception of this *nonplace*, in this section I argue that two basic aspects will characterize containment zones such as *fish tanks*, controlled areas or cleaned areas or transitional spaces: first and foremost an imposed, strong degree of security and surveillance; and second, the increased long wait times that may last hours and even full days. Within this new condition, airport officials, administrators, and eventually airport planners have undertaken several new initiatives wherein such *placelessness* becomes potentially profitable and thus *priceless*.

After the drop in the traffic indexes in the years after 9/11, the international airport system was forced to seek new sources of revenue, but the options were scarce and mainly fell into the hands of planners and architects who now needed to think of airport malls and commercial additions everywhere possible. For many scholars, "a key development in the evolution of the airport industry has been the increase in the dependence on non-aeronautical or commercial revenues" (Graham, 2001, p. 126). This is especially true in the aftermath of the tragedy and its eventual collapse, which hardly affected the industry not just in operative but in economic terms as well. During the following years and before the drop in the airlines' demand, the rule of thumb became increasing revenue's diversification at all costs. In consequence, Salter (2008) has commented that "airport architecture has moved away from the original layout of a railway terminus toward the style of a contemporary shopping mall" (p. 6), wherein "balconies, refreshment areas, and restaurants were separated from the main passenger flows" (p. 15). This strategy clearly responded to a new quantitative, statistic response from airport retail experts, wherein providing visitors and passengers with a retail offer more in accordance with their needs became less important than offering leasing space along the primary routes between check-in desk and the jetways. These measures obliged designers to *channel* passengers in order to increase retail exposure, and users were thus flow-forced to walk longer distances to their planes. The only available directionality pattern was clear: walk straight or walk back and start the processing all over again.

The wave of expansion of the security paradigm, particularly after 9/11, has shifted not just our concept of the airport but the way we understand the landside and the airside frontier. For security reasons, and perhaps for the first time ever, passengers, visitors, and merchandise were contained in hold rooms for long periods of time. This policy of containment formed a sort of undefined zone that has encouraged several hub airports to invest in retail, entertainment, hotels, and even churches, spas, and medical facilities. Finally, I concur with Salter (2008) when he argues that "the time spent on the ground at airports is not seen as 'dead-time' but rather can be quickly soaked up by the institutions of social and commercial life. In this way, airports can be seen as microcosms of society—although critical analysis on this point needs to be taken further" (p. 12).

Placeless spaces are certainly priceless. Airport retail experts Rowley and Slack (1999) have shown studies in which the "seeming placelessness and timelessness of airports stimulates consumer spending"

(p. 151). But it is in reality the idle time that has been constantly extended between check-in and departure time, which needs to be filled in. In terms of retail, "security is an indirect benefit: passengers, airlines, and terminal retailers are chiefly interested in the processing time rather than security outcomes" (Salter, 2008, p. 11).

This may well be one indirect stimuli product of slowing down passenger flows on behalf of security processing. This is what is called airport "'dwell-time,' or enforced waiting periods to create retail opportunities" (Lloyd, 2003, p. 94, as quoted in Salter, 2008, p. 11). In contrast to retail layouts of the 1970s and 1980s, according to which most if not all the leasable space was placed at the concourse, new retail schemes took advantage of passenger containment and flooded cleared areas of the terminal. At present, "shops and catering, if any, on the landside require a detour and an effort by the passenger to leave the normal flow routes" (Doganis, 2004, p. 139). This is logical if check-in time must be sped up to reach the waiting deadlines imposed by the airlines, wherein passengers must be at the gate at a certain time that is handwritten on the boarding passes.

Today most passengers will try to reduce the uncertainties and congestion of security filters, spending the least possible time at the concourse. In this regard, Anne Graham (2001), an expert on airport management, has said,

> admittedly it is difficult to determine entirely whether the raised expectations at airports have been caused by a genuine need or desire of the consumers for expanded facilities or whether an airport's drive to maximize its commercial income by becoming a shopping centre has merely changed the expectations of passengers. It is also true to say that this increased emphasis on commercial facilities has not been welcomed by all the traveling public with significant groups of passengers, particularly those from the business community, often desiring a quick route through the airport as uncluttered as possible from the distraction of numerous shops and catering outlets. (p. 127)

In regard to the new retail schemes used by architects, Adey (2008) opines that "it is important that these new layouts are not *felt* in a disruptive way or become an annoyance" (p. 149), although this is hardly possible in practice because most retail additions and reconfigurations needed to literally be squeezed in existent terminals worldwide. At least in spirit, the "principle is to avoid the psychological and physical separation between airport processing time and shopping/refreshment time" (Doganis, 2004, p. 140), but in general this was not achieved in most airports. So at the turn of the twentieth century, newer terminal retail layouts became sandwiched between landside–airside boundaries and started to look suspiciously similar to large, anchor department stores.

However, the newest developments in controlled zones indicate that the construction of a sort of micro-world, as described by Turkle (1995), is unavoidable, regardless of why passengers are staying longer and longer within the cleared areas of airports. This means that there seems to be a new, even larger opportunity for entrepreneurship at the limbo. The so-called no-man's land of the airport is gaining popularity for many reasons: Passengers may find now local shops, outlets, therapy and spas, cinemas, hotels with swimming pools, art galleries, and gardens. All of this activity is happening between the secured filters. Although I do find fascinating the idea of stretching the limbo to the maximum, it is striking to think that users are very aware that they are spending time in one of the most heavily guarded places there is. Do the implicit connotations of spending leisure time under invasive surveillance indicate the possible reassignment of personal privacy? Is this voluntary, or subtly enforced by the state?

In conclusion, I suggest that complex, historical socio-technical conditions have shaped the construction of this new undefined, unclaimed zone of uncertainty, which I call metaphorically across the chapters "the airport limbo." Finally, the investigator is tempted to speculate on whether other technological systems also have their own mediating entities, and if these exist, whether they resemble a limbo. The answer might be hidden in the three-dimensional space of computer virtuality, the sounds produced by DJs, automobile cabins, chat-rooms, or PDA's (or yesterday cellphones). Do we stand before a new technological dimension that has not yet been classified?

5 Intangible Aspects: The Immaterial Frontier

Non-physical boundaries have existed since the early days of airports. As I have shown in the previous chapters, the simplest devices that were used to separate people from aircraft created a psychological divide. For example, in LaGuardia's airport, visitors, enthusiasts, and families were not just physically but emotionally distanced from the spectacle of flight. In the cases of Washington and Tampa—while the boundary was not yet subject to excessive security controls—designers relied much more on physical or material solutions. In both cases, the autonomy for transiting between the landside and the airside has been removed. In Dulles Airport, passengers were drawn away from the apron, then isolated and contained; in the case of Tampa's airport, and thanks to the People Mover, they became fully reliant on machines for transiting between the landside and the airside. Hence the immaterial frontier may be used as a category of historical analysis. But for the purpose of this section it is particularly interesting to see the exponential growth of non-physical limits right after the tragic events of 9/11. Thinking of landside–airside boundaries in the aftermath of the tragedy and beyond presented new planning challenges, such as when designers had to address the fact that a good chunk of the frontier has dematerialized or vanished. And it is precisely that—the added uncertainties of airport technology and design—which I analyze here. The phenomenon of virtualization became a true game-changer for airport planners and designers on a global scale; nowadays, beyond the previous difficulties of laying out buildings and systems, these professionals need to understand, organize, and choreograph a new set of limits that were pretty much invisible. I argue, then, that it is nearly impossible for us to understand the current complexity of the landside–airside boundary without at least performing a taxonomical analysis. In this sense, I argue that the *virtual* manifestations of the landside–airside in the post-9/11 world could be catalogued as *psychological* constructs, informatics assets (pieces of data), and non-physical mechanisms of control.

Imagining the borderline as a psychological construct refers to how users experience a variety of feelings throughout airport terminals, depending on the material representation of boundaries and the message that designers, architects, planners, the airport itself, or often a higher authority meant to express or enforce. In this regard, both the perception and mindsets of users are directly interconnected with the ways airports are planned at the outset. As the borderline becomes a cultural limit, the landside–airside boundary transforms into a neutral zone, where cultural expressions other than the culture of airports are suppressed. As J. G. Ballard (1997), the visionary British fiction author, has said, "airports have become a new kind of discontinuous city, whose vast populations, measured by annual passenger throughputs, are entirely transient, purposeful, and for the most part happy. Above all, airports are places for good news" (p. 26–29). On the one hand, this is a true description indeed: we all have either greeted the arrival of a loved one or experienced a warm welcome from our significant others. But on the other hand, far from the optimism displayed by Ballard, contemporary "airports are sites of discipline, where passengers become passive subjects and bearers of the power relations that force compliance" (Bennett, 2008, p. 68). On a psychological level, airport terminals seem to work as large-scale thresholds of change. As passengers, we are obliged to deconstruct our persona in separate pieces, or "vectors of flow" (Salter, 2003), and they will remain split apart until passing through the airport.

Few would deny that airports post-9/11 are generally characterized by inconvenience and stress. For example, in *Air Rage: Crisis in the Skies*, the authors describe how the difficulties of expansion have resulted in "traffic jams, lack of adequate parking facilities, long lines at counters, crowded gate areas, poor restroom quality, and countless other inconveniences for travelers at the nation's airports" (Anonymous & Thomas, 2001, p. 32). The sole experiencing of these conditions could change the mood of users, even without the added stress of living in urban contexts where traffic, passing through dangerous zones to reach airports, rushes in general.

In addition, for many, arriving at the airport means that the stress path goes on. Salter (2006) has said that "traveling through the airport has become an adventure within itself—and the passenger hasn't even left the ground. A micro politics of the real physical architecture of the airport within which people spend long hours and make arduous journeys" (p. 167). But in this century, traveling by plane is a considerable odyssey involving physical and mental stoicism. Users are generally stressed upon arrival to the terminal, mentally biased to suffer mistreatments, long lines, and endless queues, lose privacy, and walk endless distances. Terminals are institutions of emotional transfer, and the landside–airside its mechanism of coercion.

For users, the emotional condition that is exalted at the airport is a game of opposites. Although our experiencing of terminals may range from joy to anxiety, from sentimental containment to emotional amplification, or from awe to rage, it is perhaps "how feelings, sensations, and emotions are sorted out and filtered by the airport. I suggest that particular modes of feeling are limited and curtailed in an attempt to both profit from passengers while also making them be, and feel, more secure" (Adey, 2008, p. 150). But for others is the "lacking of understanding of just how these sites are experienced, not just by scholars and academics but by those who do so on an everyday occurrence" (Crang, 2002, p. 569). Architects and interior designers play a key role in shaping space in airport terminals. In spite of the complex programmatic interrelationships, successful designers sort out obstacles and create clear, airy, and often monumental buildings. Because they are singular in many ways, airport terminals are often surprising and unexpected; they feature their own level of expressiveness, and "by designing atmospheric and architectural triggers that allow the drama" (Edwards, 2005, p. 74), designers transform public spaces and add monumentality. However, not all emotions are the product of experiencing spectacular buildings; architects often advised by retail experts are capable of "remodulat[ing] passengers from a stressful, anxious, and panicked state into a more relaxed mood, susceptible to consumer temptation" (Adey, 2008, p. 152). This sort of emotional fine-tuning has become one of the critical, most debated aspects of defining landside–airside frontiers.

At the opposite end, airports can also be stressful places, labyrinthine, or simply hard to navigate, and in the worst case passengers may feel intimidated. In this sense, I suggest out of the many different factors that have turned the airport into a fearful place is the extended idea that threatening and potentially dangerous people may infiltrate airport security. In this sense, fear can be experienced from two angles: the first is the fear in the face of scrutiny, and the second, racial/cultural profiling—the fear of the victim. "In the post-Cold War world, the most important distinctions amongst peoples are not ideological, political or economic. They are cultural … People define themselves in terms of ancestry, religion, language, history, values, customs and institutions. They identify with cultural groups: tribes, ethnic groups, religious communities, nations and, at the broadest level, civilizations" (Bain, 2006, p. 139). But the sentiments of public fear in contemporary travels are directly linked to the leading security paradigms and "respond to questions of cultural diversity within states" (Commission on Human Security of the Government of Japan, 2003). In practice, the more accentuated moment of anxiety is perhaps when passengers start queuing for passport control examination and face governmental controls of personal identity. These "moments of departure anxiety [are] aggravated by the sense of being funneled up to another chute like a herd of cattle" (Gordon, 2004, p. 234). It is right in that moment of anticipation that the situation becomes usually tense. A situation that is artificially stressed dramatically and almost staged publicly: A loudspeaker is cyclically repeating messages, announcing alerts, and the everyday situation is classified as a state of emergency: yellow, orange, red. Salter (2003) has referred to the design of the space of the passport control area, saying that it "parallels the legal and psychic journey from being controlled and contained at the border to the comparative freedom of the safe domestic space" (p. 124). This is often an illusion.

The fact is that the landside–airside boundary is designed through a series of valves, and some of them remained constantly opened. It is remarkable that the reliability of this system rests on the assumption of scrutiny and strong surveillance, on the one hand, and the self-disciplined, almost docile conduct of users, on the other. At the passport control areas, there are no doors, physical limits, or anything that tell us implicitly that we cannot pass through; on the contrary, they remain mysteriously open, thus facilitating somehow the flow of people and confirming that the "default spatial assumption is that most travelers will enter the territory with a minimum of delay" (Blow, 1996, p. 142).

Just as in Foucauldian discourse language expresses power relationships, landside–airside boundaries are *charged* with a martial subtext. For customs and border control officers, "to be a guard is to know all power and control" (Peterson, 1996), and that aura is usually translated as a military imposition. In all security points, but particularly in the passport control room, immigration officers display power, showcasing military training looks and weapons, showing no emotions or compassion, and interrogating people behind thick glass. "There is common structure to passport controls (facilitated by design companies and architects) in which an entry hall is divided into lines (often drawn onto the floor) that lead to inspection booths" (Edwards, 1998, Ch. 9). So in fact architectural gestures have not just organizing power but also complete theatrical scenes, human postures, and attitudes. Compared with the rail station "the airport, rather than being comfortably numbing, is a fearful place, while the aircraft itself has become a sanctuary, a blessed relief from the chaotic deluge" (Pascoe, 2001, p. 237). However, in the airport the representation of authority or power is the not only source of psychological distress. Terminals, airports, and aircraft are becoming more prone to witnessing extreme psychological responses that are even studied and classified psychiatrically, as in air-rage syndrome. "Abnormal, aberrant, and abusive behavior on the part of thousands of airline passengers each year—known as air rage, is by far today's greatest threat to safety and security of the flying public" (Anonymous & Thomas, 2001, p. 11). Air rage is usually fueled by multiple factors such as excessive frustration, alcohol or drug consumption, mental illness, nicotine craves, claustrophobia, or fear. The institutional reasons that allow air rage chiefly "include the lack of a realistic definition of air rage ... and the lack of consequences for the vast majority of air rage offenders" (Anonymous & Thomas, 2001, p. 155). In addition, the definition does not distinguish between the different social groups; "passengers who commit air rage fall into every category of air traveler: male and female; young and old; first class, business class, and economy class" (Anonymous & Thomas, 2001, p. 155). Statistically, most of these critical cases happen right at the crossing of airport boundaries or during air flight.

The next virtual manifestation of the landside–airside in the post-9/11 world is the rise of informatics assets. Succinctly, this phenomenon represents the dematerialization of the frontier into bits and pieces of information and, incidentally, a new added complexity to the conception of landside–airside boundaries, especially in large, international airports. In this section, I am interested in showing how, after 9/11, the airport became a huge "data filter" and why it is now interconnected to the larger national and international systems of personal identification.

These new policies greatly complicated the changing airport "status" of travelers according to their particular position on the physical grounds of the airport; even the coinage of new terms became necessary for this purpose (cleared, pre-cleared, sterile, non-sterile, processing, hold, metal processing, profiling, etc.). As sociologist Manuel Castells (2000) said, "data-flows are essential to emerging modes of organization" (p. 31). As ground and air transportation catch on technologically at the turn of the twentieth century, both industries started occupying a niche within the information society. More specifically, the so-called information revolution was used in a timely way to fill in the gap created by the human and technological errors of 2001. In the new reality of the information society, airports had to be rethought as *data filters*; travelers were now sorted by appearance and "scanned like barcodes" (Adey, 2002, p.13). Those with *dubious* looks or backgrounds would now take longer to be processed, and those on watch lists not allowed to travel. The new, contradictory instruction is "to maximize but to regulate mobility" (Lyon, 2003, p. 13).

After 9/11, the desperate need to reinforce airports as borders and to establish a more rigid passenger control was rapidly transformed into bytes and pieces of information; data became embedded in documents until the *pulverization* of information became the global standard. PAX, or post 9/11 travelers, became binomial to information as their individuality is now attached to passports, visas, and printed documents, which at the same time connected them to a local or international electronic database. At the light of high security and the airport world after 9/11, the landside–airside boundary became the ultimate international frontier. These visible/invisible *edges* would now interplay with national identity cards, stretching international borders into terminals and making them increasingly immaterial.

The third virtual manifestation of the landside–airside in the post-9/11 world is the non-physical mechanisms of control. One of the more severe consequences of the 9/11 aftermath is that the exercise of state control is now stretched to a limit, and in doing so, governments are getting dangerously close to violating privacy, free transit, and other basic civil rights. Under this new paradigm, the landside–airside boundary could be thought of as partly a mechanism of non-physical control that is contingent on international regulations on one side and on local practices and culture on the other. According to Orvar Löfgren (1999), an expert on national borders and a professor of enthology in Sweden, the airport is "a surveillance machine for automated bodies, shepherded from control station to control state" (p. 17). In many ways, the central concern of the literature around airports, especially the attending problems of international relationships, government, and law, has focused on the growing relevance of airports as tools of societal control and surveillance.

If contemporary airports mediate the relationships between citizens and the state, how are these non-physical mechanisms of control contingent on a reformulation of the boundary? The resulting policies of risk management and threat mitigation that were imposed by the NSA on the ICAO and IATA resulted in a domino effect on the airports worldwide. Beyond following the new code emends that demanded new criteria for buildings, technologies, and procedures, designers faced the added challenge of dealing with non-physical strategies of control. But, as in all emergencies, some of these mechanisms or strategies were written and some were not, a; though most were learned with practice and use. Among the frequently used non-verbal actions are the idea of permeating a sense of permanent risk (*collective paranoia*); the availability of everyone for ID checks (*scannability*); the imposition of a passive and docile attitude, especially on the part of passengers (*passivity*); the widespread notion that force will be exercised against suspicion (*restrictiveness*); the use of learned routines and patterns of conduct, movement, reaction, and so forth (*mechanization*); and, finally, the preparing of civilians to react to and intervene in potential threats (*alertness*).

Thus, I argue that non-physical mechanisms of control are powerful links between the practice of airport design and the larger discourses behind politics, economics, or the state. This condition may allow designers to observe and register how users adapt or react to the new ideology. These are those tactical measures that superseded political measures—often shielded in national security and border security—and that have seriously affected the way landside–airside boundaries were conceived after 9/11.

Particularly in the USA and Europe, institutions spread a sort of collective paranoia wherein threats seem to be omnipresent. "Risk is then, mitigated, avoided, transferred, or accepted according to the abilities and environment of the authority. Risk management is the identification, analysis and elimination (and/or control to an acceptable level) of those hazards, as well as the subsequent risks that threaten the viability of an organization," according to ICAO (2005, Section 5:3). Potential risk is present in any landside–airside boundary design because each material, surface, or route must be thought out in terms of a worst-case scenario. Indeed, threat and risk management is one of the governance keystones of our time.

After 9/11, measures were implemented to render people scannable. Salter (2003) has said that "airports are perhaps the most stringently surveilled sites in terms of the means of movement and of identification" (p. 34). But in terms of control, the Orwellian eye behind the mirror is the pure representation of institutional superiority and mental voyeurism. In practice, everyone must show sufficient identification, and as airports become cultural limits, racial and religious profiling becomes an edgy aspect of the virtual borderline. Federal and local authorities have imposed a written/non-written conduct code that implies important degrees of passiveness from the user's end. Passengers are restricted from taking photos, making jokes, using their cell phones, talking loudly, or even sitting down to rest. For Salter (2003), "the sovereign, taxing power of the state and the docile, self-policing nature of the citizen are reinforced immediately" (p. 127); but on the other side, the state imposition of emotional control implies a sort of new designed or planned ideological social order, like the one James C. Scott discusses in his book *Seeing Like a State* (1998).

There is a widespread notion, be it true or false, that the airport is an incredibly restrictive place. "Airports are sites of discipline, where passengers become passive subjects and bearers of the power relations that force compliance" (Bennett, 2008, p. 68). Governments are represented in many ways in an airport, but basically they have physical presence in order to exercise punitive actions or control the population. Thus, the intersection between the entities involved, be they users, authorities, or technological artifacts, is normalized by written and non-written codes that go beyond the Law in cities, for example, and potentially the use of force (weapons, physical force, strategy, etc.) will be used against for compliance.

Similar to the regulatory features implemented by modern states in order to control population discussed in *Seeing Like a State* (Scott, 1998), contemporary institutions have choreographed—through highly technified designs—mechanized patterns of movement. Passengers have learned to perform certain routines in an automatic fashion, as in the case of a six-year-old girl who reacts to an inspection, performing and behaving like an adult (IRBMW, 2011). In regard to how contemporary landside–airside boundaries are thought of, confinement operates at subconscious levels, especially in the immigration hall and customs, where no physical barrier seems to obstruct the flow of people.

Post-9/11 immigration halls are governed more by attitudes and signage than by architectural gestures or imposing spaces. In the end, everyone at the airport must remain alert. Many routines performed by the airline and security personnel, signage, posters, video projections, and, particularly, announcements through loudspeakers have as their mission to remind all visitors and passengers to remain attentive and distrustful of strangers. The state goal was clearly to expand its presence and vigilant force to the side of civil society, but in doing so it strengthens the imposed ideological order that, at least today, governs the airport.

Finally, beyond the control policies of the state, it is the growing interest of governments and other international regulatory bodies that these "Checkpoint Charlies" of today share information, allow interagency controls, impose arbitrary policies, divide nations, cultures, and even religions, because we seem to be entering into the "*transnationalization* of bureaucracies of surveillance" (Bigo & Tsoukala, 2008, p. 14).

6 Boundaries and Beyond

Airports have served many purposes other than being huge parking bays for airplanes. In the past, airports such as LaGuardia or Tempelhof became vehicles of political propaganda; during the early-to-mid-twentieth century Lambert Field and TWA became the new urban cathedrals, and Orly revealed itself as a monument to modernism and modernity. A decade later, airports became huge water tank-like containers, as at Charles de Gaulle T1 and Tampa International. After the first terrorist attacks in the early 1970s, Ben Gurion became a living fortress and the archetypical airport-bunker; in the following decades Heathrow and Schiphol evolved into air-conditioned malls. The post-9/11 world turned Atlanta-Hartsfield and other hubs into sort of martial "Safety States" (Lyon, 2008, p. 30), while the contemporary Singapore's Changi accommodated parks, fountains, swimming pools, hotels, and entertainment within the landside–airside stretch; Bogota's El Dorado airport has its own TV show about drug smuggling, Hong Kong's Chek Lap kok airport has an Imax Theatre in Terminal 2 or even Seoul's Incheon airport its own zen garden!

The next airport will perhaps explore and exploit more intensely what I have called here the limbo, the in-between, or the contained zones in the landside–airside boundary. I would not hesitate to say that even other kinds of viewports might be found in other technological systems. With its high degrees of personal control, tax-free policies, jurisdictional vagueness, target-consumers' buying power, and consumers' insatiable thirst for retail and entertainment, and while operators can keep them secured as prisons, the limbo will probably remain the favorite spot of governments and companies. And although seeing what is next is not the purpose of this work, the paradigm of security represents a continuum that will take many years before it is subject to any changes that may ease the original post-9/11 policies. It is thus foreseeable that identification control and passenger scanning and people containment will remain as game constants; thus it is reasonable to think that biometrics will play an important role in mediating bodies and boundaries.

However in the coming years, the state will be perhaps more interested in biometric control and the resulting databases than in airport security. It is not ludicrous to think of airport terminals as fully *clean zones*, where the ID and security screening has been pushed outwards, toward the incoming roads. In the near future, passengers may be sorted using scanning devices that track body features, way before they reach the terminal building. Throughout this chapter, I have used the metaphor of *liquid airport* because the designing of airport terminals since the rise of security issues in the early 1970s has become, more and more, an activity of flow containment. As I have shown in the previous sections, passenger contention and flow distribution are achieved through a complex set of devices, decisions, technologies, walls and partitions, routines, and invisible mechanisms that I have defined in general as *viewports*, or the boundary between the landside and the airside.

But there is also the issue of how transparent airport systems have been, are today and will be in the upcoming future. I personally like to think of it as a historical security gradient that will always shift according to political convenience and paranoia. How transparent airports are is the matter of a very intriguing study by Rachel Hall from Louisiana State University. Hall argues that the "aesthetics of transparency is at work in security cultures ranging from US-controlled airports to those not under US control, from war zones to occupied territories, from professional police work to neighborhood watch programs, and from No-Joking Zones in airplanes to analogous controlled-time zones beyond airports. What unites each of these geographically, culturally, and politically diverse security cultures is a paranoid orientation..." (Hall, 2015, p. 157). Indeed, it is an intangible fear to others what drives policy behind global airport systems. If airport security became at one given moment a serious architectural challenge, nowadays is better portrait as a technological one; however, I foresee that in the future the concept of security will move on and transform into a more radical sociopolitical form of control. The utopia of globalization is doomed to face the opposite, strong movements of reborn nationalisms and increasing social tribalization. Airports will be at the epicenter of an expanded identity crisis, where only those *wanted* or *tested* will travel. In an inverse proportion, just as the world information networks expand, the conveniences of flying and entering into other countries may become gradually more restrictive. It would not be crazy to speculate that the *virtual persona* may be able to travel universally (into the information world), although the same *physical persona* may be heavily restricted to leave even from their own country. It may well be a new, variation of the grim scenarios predicted by Stanislav Lem or George Orwell. The ideals for a free, democratic, peaceful, and united world are dying in front of our eyes; as the rise of a new breed of ignorant and megalomaniac dictators take over the Western hemisphere and beyond, the impossibility of free flow is ahead. Migration will see an end and countries will eventually shell out. Traveling in the future may be soundly dangerous depending on who you are, how you look or what you think and the airport will become the perfect filter to reject you, or even put you in prison.

References

Publications

Adey, P. (2002). *Secured and sorted mobilities: Examples from the airport. Institute of geography and earth sciences*. Aberystwyth, UK: University of Wales.

Adey, P. (2008). Mobilities and modulations: The airport as a difference machine. In M. B. Salter (Ed.), *Politics at the airport*. Minneapolis, MN and London: University of Minnesota Press.

Altshuler, A., & Luberoff, D. (2003). *Mega-projects: The changing politics of urban public investment*. Harrisonburg, VA: Lincoln Institute of Land Policy.

Anderson, B. (1983). *Imagined communities: Reflections on the origin and spread of nationalism*. New York, NY: Verso.

Anonymous & Thomas, A. R. (2001). *Air rage: Crisis in the skies*. New York, NY: Prometheus Books.

Bain, W. (Ed.). (2006). *The empire of security and the safety of the people*. Oxfordshire, UK: Taylor & Francis Group.

Ballard, J. G. (1997). Airports: The cities of the future. *Blueprint, 142.*

Bennett, C. (2008). *The privacy advocates: Resisting the spread of surveillance.* Cambridge, MA: The MIT Press.

Bigo, D. (2004). Globalization of In-security: The field of professionals of unease management and the ban-opticon. *Traces: A Miltilingual Journal of Cultural Theory, 4*, 109–112.

Bigo, D., & Tsoukala, A. (2008). *Terror, insecurity and liberty: Illiberal practices of liberal regimes after 9/11.* New York, NY: Routledge.

Blow, C. J. (1996). *Airport terminals* (2nd ed.). Oxford, UK: Butterworth-Heinemann.

Brancker, J. W. S. (1977). *IATA and what it does.* Netherlands: A. W. Sijthoff.

Castells, M. (2000). *The rise of the network society.* Malden, MA: Blackwell.

Commission on Human Security of the Government of Japan. (2003, May). *Human security now: Protecting and empowering people.* Clash of civilizations.

Crang, M. (2002). Between places: Producing hubs, flows, and networks. *Environment and Planning A, 34*(4), 569–574.

Curnow, A. (1997). Narita. In *Early days yet: New and collected poems 1941–1997* (p. 32). Manchester, UK: Carcanet.

De Neufville, R. (1975). Airport systems planning. In R. Horonjeff (Ed.), *Planning and design of airports* (2nd ed.). New York: McGraw-Hill.

Doganis, R. (2004). *The airport business.* New York: Metropolitan Books.

Douglas, D. (1995). Airports as systems and systems of airports. In W. M. Leary (Ed.), *Airships to airbus: The history of civil and commercial aviation.* Washington, DC: Smithsonian Institution.

Edwards, B. (1998). *The modern terminal: New approaches to airport architecture* (Ch. 9). New York, NY: Routledge.

Edwards, B. (2005). *The modern airport terminal: New approaches to airport architecture* (2nd ed.). New York, NY: Spoon Press.

Foucault, M. (1977). *Discipline and punish: The birth of the prison* (A. Sheridan, Trans.). New York, NY: Vintage Books, Random House.

Fuller, G., & Harley, R. (2004). Anatomy of an airport. In *Aviapolis: A book about airports.* London, UK: Black Dog.

Gordon, A. (2004). *Naked airport.* New York, NY: Metropolitan Books.

Gottdiener, M. (2001). *Life in the air: Surviving the new culture of air travel.* New York, NY: Rowman and Littlefield.

Graham, A. (2001). *Managing airports: An international perspective.* Oxford, UK and Boston, MA: Butterworth-Heinemann.

Hall, R. (2015). *The transparent traveler.* Durham, NC and London, UK: Duke University Press.

Harmel, K., & Spadanuta, L. (2006, September 9). Mickey wants your prints. *Globe and Mail*, p. T9. Toronto.

International Civil Aviation Organization. (1974). *Annex 17 to the Convention on International Civil Aviation* (1st ed.). International Standard and Recommended Practices. Montreal, QC: Author.

International Civil Aviation Organization. (2005). *ICAO's accident prevention program manual* (Document No. 9422). Section 5:3.

IRBMW. (2011, April). *6 years old girl physical inspection (U.S. Airport).* Retrieved January 15, 2013, from http://www.youtube.com/watch?v=ApmBa-ZSh9E.

Khurana, K. C. (2009). *Aviation management, global perspectives.* New Delhi, India: Global India Publications.

Lahav, G. (2008). Mobility and border security: The U.S. aviation system, the state, and the rise of public-private partnerships. In M. B. Salter (Ed.), *Politics at the airport.* Minneapolis, MN and London: University of Minnesota Press.

Latour, B. (1993). *La Clef de Berlin et Autres Leecons d'un Amateur de Sciences.* Paris: La Decouverte.

Leggett, B. (2012). Departures: The novel, the non-place and the airport. *Alluvium, 1*(4), no page.

Ley, D., & Cybriwsky, R. (1974). Urban graffiti as territorial markers. *Annals of the Association of American Geographers, 64*(4), 491–505.

Lloyd, J. (2002). Departing sovereignty. *Borderlines e-journal, 1*(2), 1–12.

Lloyd, J. (2003). Dwelltime: Airport technology, travel, and consumption. *Space and Culture, 6*(2), 93–106.

Löfgren, O. (1999). Crossing borders: The nationalization of anxiety. *Ethnologia Scandinavica,* 29.

Lyon, D. (2007). Surveillances studies: An overview. Cambridge, UK and Malden, MA: Polity Press.

Lyon, D. (2008). Filtering flows, friends, and foes: Global surveillance. In M. B. Salter (Ed.), *Politics at the airport* (Ch. 2). Minneapolis, MN and London: University of Minnesota Press.

Pascoe, D. (2001). *Airspaces.* London: Reaktion Books.

Passi, Anssi. (1996). *Territories, boundaries and conciousness: The changing geographies of the finish-russian boundary.* New York, NY: J. Wiley and Sons.

Peterson, S. (1996, January 2). Checkpoint Charlies dash British convoy's blithe spirits. *Christian Science Monitor*, p. 4.

Reitman, I., & Reitman, J. (2009). *Up in the air [motion picture]*. USA: Dream Work Pictures.

Rowley, J., & Slack, F. (1999). The retail experience in airport departure lounges: Reaching for timelessness and placelessness. *International Marketing Review, 16*(4/5), 363–375.

Salter, M. (2003). *Rights of passage: The passport in international relationships*. London, UK: Lynne Rienner Publishers.

Salter, M. B. (2005). At the threshold of security: A theory of international borders. In E. Zureik & M. Salter (Eds.), *Global surveillance and policing: Borders, identity, security*. Cullompton, UK: Willan.

Salter, M. B. (2006). The global visa regime and the political technologies of the international self: Borders, bodies, biopolitics. *Alternatives, 31*(2), 167–189.

Salter, M. B. (2007). Governmentalities of an airport: Heterotopia and Confession, *International Political Sociology, 1*(1), 46–67. Reprinted in M. B Salter (Ed.). (2008). Introduction: Airport assemblage. *Politics at the airport*. Minneapolis, MN and London: University of Minnesota.

Salter, M. B. (Ed.). (2008). *Politics at the airport*. Minneapolis, MN and London, UK: University of Minnesota Press.

Schaberg, C. (2017). *Airportness, the nature of flight*. New York, NY: Bloomsbury Academic.

Scott, J. C. (1998). *Seeing like a state: How certain schemes to improve the human condition have failed*. New Haven, CT and London, UK: Yale University Press.

Scullin, G. (1968). *International airport: The story of Kennedy Airport and U.S. Commercial Aviation*. Boston, MA and Toronto, ON: Little, Brown.

Stanton, J., Chango, M., & Owens, J. (2008). *ICAO and the biometric RFID passport: History and analysis*. Syracuse University, School of Information Studies.

Turkle, S. (1995). *Life on the screen*. New York, NY: Touchstone, Simon & Schuster.

Van Houtum, H., Kramsch, O., & Zierhofer, W. (Eds.). (2005). *B/ordering space. Border regions series*. Burlington, VT: Ashgate.

Van Houtum, H., & van de Velde, M. (2004). The power of cross-border labour immobility. *Tijdschrift voor Economische en Sociale Geografie (TESG), 95*(1), 100–107.

Van Houtum, H., & van Naerssen, A. L. (2002). *Bordering, ordering and othering*. Amsterdam, NL: Royal Dutch Geographic Society.

Wilson, T. M., & Donnan, H. (Eds.). (1998). *Border identities: Nation and state at international frontiers*. Cambridge, UK: Cambridge University Press.

7
Conclusion

Perhaps it was after my first contact with one of the most obscure books ever written on transportation technologies, the strange, but always fascinating *Aviapolis: A Book About Airports*, by Gillian Fuller and Ross Harley (2004), that I became concerned about this topic. I use the word *concerned* because the boundary, at least in the context of the book, is ominous and grim. *Aviapolis* throws a fastball to the airport establishment of glamor and majesty thanks to its diagrammatic formatting, un-retouched photographs, and coldly descriptive texts. Toned with bare realism, most of the book follows the tracks of where things get divided in an airport, generally without explaining really why or how. But it was right then that I was provoked by the idea of coming up with a better explanation for why the landside–airside boundary is the defining element of the ways airports look and function. It was certainly at that point when I started weaving my expertise as an airport designer and my research as an S&TS scholar. The initial result of my inquiry filled me with questions, but I decided to focus on answering just two basic ones: Can we challenge the way we have written the history of airports and the form in which we understand them today and perhaps tomorrow? Can we prove that the landside–airside boundary is the

V. Marquez, *Landside | Airside*, https://doi.org/10.1007/978-981-13-3362-0_7

single most important feature that shapes an airport? The answer to the second question is yes, it is, *today*. The catch of this simple answer is that the landside–airside boundary is not necessarily an essential part of the airport. It is a socio-technical construct, though, that we have complicated until there is no other way to think of an airport *today*, except in terms of the boundary itself.

To the first question, it is much harder to provide a convincing answer. I maintain in this book that airport planning and design have moved along a *straight line* for more than a century. This pattern of *linearity* has been supported by what planners call "standard airport design," and it basically consists in learning from what is established[1] as modestly successful (airport typology) and in introducing small, superficial changes to the selected scheme (finger, satellite, pier, unit, etc.). Although the historiography of the airport is most often pictured in terms of spectacular achievements, mostly architectural and spatial, but this interpretation could be deceptive and misleading. Why? Because most airports are based on the same standard airport layouts of the past (sometimes almost a century old). Furthermore, the airside has shown few changes since the Wright brothers took flight (Stover, 2003, p. 31), thus this *linear* process of thinking has privileged the demands of aircraft (dimensions, technological features, turnaround time, etc.) and airspace (slots, frequency), leaving behind the needs and expectations of passengers. It has also adjusted the landside of the terminal mostly for the automobile.

Therefore, I have decided to introduce the idea of a landside–airside boundary as a research tool, a sort of a *lens* that in hindsight could allow us a new glimpse into the history of airports. A frontier that should be thought not as a line or as a door, but as said by the famous American filmmaker John Sayles, as a virtual or physical space that belongs neither here not there, that in the end turns out to be a metaphor of the inconclusiveness of the airport as a technological artifact:

---

[1]"*Established*" by experts and administrators, not necessarily by users.

> Limbo is basically a metaphor, anyway … it's this place where you're neither here nor there … and so many people live that kind of life.
>
> —John Sayles (2009), *Limbo, A Condition of Unknowable Outcome*

So, based in the aforementioned, can we challenge the way we have written the history of airports and the form in which we understand them today and perhaps tomorrow? In conclusion, I argue that it is in fact possible, but only through the observation and understanding of landside–airside boundary changes throughout the years. In the previous chapters, I presented testimony that shows why the airport, as a technological system, faced at least three bottlenecks throughout its history.

Each of these specific moments of crisis pushed for a sound, potent "reverse salient" that eventually could lead to a new process of invention. Cyclically, airports have fallen behind and have been phased out as *efficient* technological objects; designers have tried hard to correct the standard models, but have failed in their intent. Thus the problems (delays, congestion, passenger complaints, etc.) mounted and turned out to be so radical that the government pushed for the creation of a new, competitive system. These atypical cases, all three exceptional, may be also understood through the use of other S&TS concepts.

LaGuardia, Dulles, and Tampa shared the basic characteristics and structure of SCOT, or social constructivism, meaning that in these three cases, there is interpretive flexibility, a problem faced by relevant social groups, a process of stabilization and rhetorical closures. In addition, I have analyzed, in more detail, Dulles under Hughes's model for large technological systems and Tampa under Pinch & Bijker's model of technological frames.

LaGuardia—or the first reinvention of the airport—illustrates from one angle Scott's criticism of how the State enforces control over citizens through plans, artifacts, or policies; and from the other, how Mayor LaGuardia succeeded in reinforcing his political power while at the same time remaining in good standing with his fellow airport enthusiasts and voters. In this sense, it became a *soft transition*, and not a drastic one, as many visionaries predicted, and the Skywalk patented that assertion. LaGuardia Airport can be considered a monumental techno-political device and one of the biggest triumphs of the "American Sublime" (Nye, 1994). Its gestures and unprecedented features were indeed revolutionary at the time—such as the separation of departures and arrivals on two levels—and also set the basis for the new relationship between users and aircraft.

Dulles—or the second reinvention of the airport—which was intended to be the new seed for a super-system of airports, represented a radical departure from standard airports, although it entailed an enormous risk therein. I recounted how faith in technology at the turn of the 1960s built up the idea that the Mobile Lounge was the ultimate solution for the modern airport, and eventually a vehicle of social progress. I paid special attention to the formation of relevant groups, actors, and their interests; I tried to unveil the internal (experts) and external (boards) negotiation processes that explain why the reformulation of the airport is based solely on rethinking the landside–airside boundary. I have also made an account of the postwar period environment to plan airports, the jet-air boom, and the congestion of international airports at the turn of the 1960s. I also commented on how the establishment of the airport planning profession fueled the momentum of new airport thinking.

Tampa—or the third reinvention of the airport—brings the first opportunity for airport planners to rethink an airport from the ground up. I have used a slightly modified version of Bijker's tentative list of elements of a technological frame model. I have found along the way, particularly after analyzing the interviews, that there are three main levels to the gestation process of Brink Coridan's "Landside Airside Concept." The first may be called "problem posing," which is strictly semantic and where the relevant actors need to fix terms in order to identify and make complex problems intelligible and manageable. The second stage may be called "problem framing" and is mainly methodological. At this stage, definitions become the main element of communication between the members of the different relevant groups. The third stage could be called "problem solving."[2] Planners will team

[2]Assuming that not all actors are yet fully advocating—either the problem or the method (or individually).

up one last time and, based on the procedure's outcome data, will brainstorm an integrated solution that becomes the basis for a rhetorical closure.

Yet this is only half the story. Although I could have gone on endlessly speaking of the boundary—simply because it is not a *thing* or a *phenomenon* but a category of analytical research—I decided to write two more chapters that answer two additional key questions: Are boundaries repeatable pieces of a mechanism, or are they autonomous entities, subject to absorb cultural practices? And are there any features that become visible only through the use of a landside–airside boundary as a lens? Each question is developed and responded to separately.

I intuited that the differences between airport reinventions and airport innovations might intrigue readers. In conclusion, there is one basic difference between the two: an airport reinvention requires that a higher entity commissions a third party to develop a new concept for an airport without limitations in terms of organization, technology, or structure. However, the new airport must adequately fit within the higher system in the chain of transport as Debbie Douglas has pointed out, in this case, aviation. In the case of innovations, change is introduced more gradually, because many traces of the previous *standard* solution remain. Perhaps the more relevant finding in this regard is to observe that either strong innovations or reinventions happen only when the relationship of the landside–airside boundary is modified. As I have shown in this chapter, during a reinvention process, designers need an unusually high degree of freedom from their client. Clearly, this condition proportionally sparks multiple degrees of uncertainty and risk, thus making the reinvention of airports in the future significantly less probable. As designers gain decision control over the emerging airport system, they gradually resemble Hughes's characterization of a system builder. Finally, I ask myself if landside–airside boundaries are cultural mirrors, and the answer is yes. Airports speak much of local cultures, and these expressions are unexpectedly present in any landside–airside boundary.

Last, I respond to this inquiry: Are there any features that become visible only through the use of a landside–airside boundary as a lens? In trying to answer this simple question, I bumped into some interesting methodological speculations. First, the identification of these boundaries opens an imaginary window that may be described as a technological *viewport* (a term borrowed from the culture of computers and better explained Chapter 6). These portals allow us to see the in-between *space*, which is usually otherwise inaccessible and which marks the territorial border between us and the hardware of technological systems. Second, I have also found that in order to make visible what has remained hidden within those contained boundaries, viewports should meet some conditions. At least in the case of airports, I have found at least five features that characterize these portals. Viewports should be discursive, normative, the object of knowledge, transitory, and intangible.

As I have shown in "The Liquid Airport," passenger *contention* and flow distribution is achieved by a complex set of devices, decisions, technologies, walls and partitions, routines and invisible mechanisms that I have defined in general as the boundary between the landside and the airside. I have explained why these boundaries have been mentioned only intermittently in the available literature of airports (corporate and public), how they become evident when we use them as a category of historical analysis, or, furthermore, as a filter that helps us look for, in hindsight, a new version of why airports are the way they are now. I have also argued, perhaps in too much detail, how landside–airside boundaries are shaped in very unusual forms and out of a very wide array of factors. They also can be considered cultural mirrors, because each boundary, regardless of its degree of achievement, is a living version of a diversity of large-scale conditions such as the political or economic discourse of the time, nationalism or the greater concerns of society, and so forth.

In the previous investigation and analysis, I have also suggested that it is only through the meticulous pondering of the many events that take place between the conceptualization and the operation of boundaries that they become identifiable, intelligible, and modifiable. In this regard, I have addressed multiple material aspects such as innovation, geometric configurations, congestion, cultural idiosyncrasy, security, codes, literature, retail; as well as non-material aspects such as psychological constructs, informatics, the state control over users, and so forth.

The analysis made for this last chapter may be situated right at the intersection where the landside–airside is negotiated; I opted for this strategy because each of the subsections represents in itself a specific window to observe and understand how the boundary is finally formed.

But, in the end, so much complexity pays off, because the invisible becomes visible and the landside–airside boundary finally rises as the seed that gives genesis to all airports. I hope this book will help readers and airport users—that like me are often intrigued by the peculiarities of airports, to understand the reasons behind those features that are often so painful and those that make them so special.

References

Publications

Fuller, G., & Harley, R. (2004). Anatomy of an airport. In *Aviapolis: A book about airports.* London, UK: Black Dog.

Nye, D. (1994). *American technological sublime.* Cambridge, MA: MIT Press.

Sayles, J. (Director). (2009). *Limbo, a condition of unknowable outcome* [Film]. New York, NY: Green/Renzi.

Stover, D. (2003, March). Airport of the future: Science Newsfront. *Popular Science.*

Postscript

As a short postscript, I would like to end this book with a summary and evaluation of my own use of S&TS concepts and methods in order to unthread and understand this complex topic. In general, the work throughout the chapters is framed by social constructivism, meaning that I do not take for granted the notion that airports are the result of technological change, such as in aircraft, systems, and security. Starting off from that position I have investigated why we have not built an alternate history wherein social forces are rightly credited. In parallel, I have identified what I have called "standard airport design," a simple but effective and highly technologically deterministic[1] idea that has eased the way aircraft operate at the airside, somehow subordinating the airport to aviation and, in a secondary role subordinating the landside to the requirements of the automobile. Designers and planners have lived comfortably within this superstructure, producing refried versions

[1] I claim that "standardairport design" is simply a technologically deterministic formula because it favors and sometimes forces airport layouts, and more specifically landside-airside boundaries to accommodate technologies such as aircraft, security and systems (baggage handling, etc.). Despite several efforts against, the human relationship with the airport has been mostly driven and determined by technological forces.

V. Marquez, *Landside | Airside*, https://doi.org/10.1007/978-981-13-3362-0

of old models, cloning most airports in essence but wrapping some of them in gold. Nonetheless, right at the underbelly of the airport, things have started to heat up since the late 1930s, when contingency showed up and passengers started to fight back, congestion turned airports into time bombs, and traveling became a painful adventure. These "standard airport designs" may be seen as a linear path of development, wherein some irregularities (social responses) have protruded, forming what Hughes defines as "reverse salients." These super-scale technological bugs have impeded the triumph of the airport as a machine. That is basically how I realize that those three lonely trials of reinventing the airport from scratch were immensely relevant. It is impossible to visualize and understand the caliber of these cases without Hughes's model.

Once these reinventions were mapped, other meaningful innovations became visible and interconnected. All together formed a sort of anti-establishment network linked by strong changes in the way the frontier between the landside and the airside was thought of. My initial reaction was to see these frontiers as a variation of Star and Griesemer's (1989, p. 393) "boundary objects." My disappointment came later in my inquiry, because I found that the concept of the landside–airsideboundary was not necessarily immutable or integral among expert communities. So I had to come up with my own terms and define landside–airside boundaries as technological *viewports*. These entities are described further in Chapter 6.

Throughout the body of the text I make specific use of three ideas, each of which supports the analysis of one case study and thus one chapter. For LaGuardia Airport I based the construction on the chapter thinking of Nye's (1994) idea of the technological sublime, because regardless of the success or failure of the airport, it instantly became a true model of American selfhood. In the case of Dulles Airport, I used Hughes's model of large technological systems to show how airports form systems, and how they gained momentum between the 1950s and 1960s. I also discussed the coincidences between the system builder and Eero Saarinen. In other sections, I use two concepts from Madeleine Akrich (1992), first her idea of "heterogeneous technical object," which I apply in the case of the invention of jetways, and second, "imprinted objects," for studying the implications of finding routines and patterns in the Mobile Lounge.

Last, in the case of Tampa's airport, I used Pinch and Bijker's theoretical concept of the technological frame to explain why these processes comprise social, material, and cognitive elements as well as to understand the interactions among the relevant social groups. Finally, in SCOT's terms, I argued that in all three reinventions of the airport, authors reached only a rhetorical closure. This idea has been incredibly valuable to my understanding of how the relevant social groups have "accepted" the new airport layouts; and why until the adoption process has been stabilized, the birth of a radically new airport is pompously and proudly announced (applicable to all three cases).

In all chapters, I have also paraphrased some of Scott's ideas regarding how the vertical powers of the state and other institutions prostrate social interests—as in the case of many of the ideas that govern the mentalities of airport planning. Chapter 5 shows the differences between reinvention and innovation, much in the spirit of Rogers (1962) and Hughes (1983a). In the second part, I bring out the cultural implications of designing landside–airside boundaries, which is a way to demonstrate how these entities are resistant to standardization. In Chapter 6, I propose the idea of *technological viewports* and structure the way this theoretical concept can be identified and consolidated as autonomous pieces of knowledge.

Finally, in hindsight, I hope that the identification of what I have called tentatively viewports may help other researchers that are having a hard time in explaining technological change. In this regard, my suggestion is to look for certain conditions that, at least in this case, should be met. In my opinion, these spaces exist only in-between users and technological systems, but not necessarily at the operable interfaces or hardware controls of equipment. Viewports are not "operational," thus the number of cases may be significantly reduced compared to either those which remain without closure, or those that are creating "limbo" zones. Of the first examples that come to my mind is the virtual reality space in computers, which is surprisingly similar to the airport's limbo. It in fact meets all the conditions or features that I mentioned before: It is discursive (I think of Paul Edwards's *The Closed World*), normative (Internet Privacy Act), an object of knowledge (a vast body of literature is available), transitory (by nature it is a through space), and, last, intangible (although it is mostly virtual, it is already modifying our real world). The algorithmic environment of the computer is indeed an "in-between space" that needs portals of access and egress; it also plays with its own set of rules, and in the end it also challenges the legal system and awakes a compulsive need for control and surveillance. Just like in the airport boundary, the virtual reality's space gives the illusion of independency and autonomy. These viewports are spaces that have been paused, because they will never reach closure and probably will forever remain unresolved. Strangely, most probably we will be metaphorically "dwelling" inside these viewports at an increased rate that we may not be able to imagine, and we are already spending much of our time and money just there. It is intriguing to think that the faster technologies are pushed forward they tend to threat the larger frameworks of society, such as laws and norms, moral and ethical codes, religious principles, organizational and corporate standards, and so on. When this happens and no one is able to "frame" the new system within the old, viewports seem to appear. Now we can learn that in spite of their transient nature, they can grow and become permanent. I would suggest that because of their unique properties—specially its *placelessness*, we will see more and more of this thresholds cleverly escaping from the tight controls of larger entities, such as the state itself.

References

Publications

Akrich, M. (1992). The de-scription of technical objects. In W. Bjiker & J. Law (Eds.), *Shaping technologies.* Cambridge, MA: MIT Press.

Hughes, T. P. (1983a). *The evolution of large technological systems: The social construction of technological systems*. Cambridge, MA: MIT Press.

Nye, D. (1994). *American technological sublime.* Cambridge, MA: MIT Press.

Rogers, E. M. (1962). *Diffusion of innovations* (4th ed.). New York, NY: Free Press.

Star, S., & Griesemer, J. (1989). Institutional ecology, "translations" and boundary objects: Amateurs and professionals in Berkeley's Museum of Vertebrate Zoology, 1907–39. *Social Studies of Science, 19*(3), 387–420.

Bibliography

Interviews

Brink Coridan, M. (2008, March 8). Transcript. *Personal interview with Marge Brink Coridan with Leigh Fisher & Associates.*

Dunlay, W. J. (2008, March 8). Transcript. *Personal interview with Bill Dunlay with Leigh Fisher & Associates.*

Lewis, E. (2008a, March 7). Transcript, Part 1. *Personal interview with Gene Lewis with Leigh Fisher & Associates.*

Lewis, E. (2008b, March 9). Transcript, Part 2. *Personal interview with Gene Lewis with Leigh Fisher & Associates.*

Sherret, A. (2008, March 8). Transcript. *Personal interview with Alistair Sherret with Leigh Fisher & Associates.*

Publications

Activities are channelized and separated: Washington National Airport. (1941). *Architectural Record, 90*(1), 53.

Adey, P. (2002). *Secured and sorted mobilities: Examples from the airport. Institute of geography and earth sciences*. Aberystwyth, UK: University of Wales.

V. Marquez, *Landside | Airside*, https://doi.org/10.1007/978-981-13-3362-0

Adey, P. (2008). Mobilities and modulations: The airport as a difference machine. In M. B. Salter (Ed.), *Politics at the airport.* Minneapolis, MN and London: University of Minnesota Press.
Aéroports de Paris. (2004). *Roissy—Charles de Gaulle: 30 ans de déceloppement économique et de partenariat local.* Paris, France: Author.
Aéroports de Paris. (2009). *Terminal 1 at Paris—Charles de Gaulle airport celebrates 35 years and adopts a new image.* Paris, France: Author.
After Tampa, airports will never be the same. (1971, April 11). Advertisement from Westinghouse. *Tampa Tribune–Times.*
Air Canada traffic grows. (1971, April 11). *Tampa Tribune–Times.*
Air fuel goes underground. (1971, April 11). *Tampa Tribune–Times.*
Airport bus would haul passengers right to plane. (1953, July). *Popular Science.*
Airport police prepared, just in case. (1971, April 11). *Tampa Tribune–Times.*
Airports in Latin America. (1930). *Trade Information Bulletin No. 696.* Washington, DC: U.S. Department of Commerce.
The airport scramble. (1956, June). *Architectural Forum.*
Akrich, M. (1992). The de-scription of technical objects. In W. Bjiker & J. Law (Eds.), *Shaping technologies.* Cambridge, MA: MIT Press.
Altshuler, A., & Luberoff, D. (2003). *Mega-projects: The changing politics of urban public investment.* Harrisonburg, VA: Lincoln Institute of Land Policy.
Ammann & Whitney, Eero Saarinen, Burns & McDonnell, Ellery Husted, & Burhan Kelly. (1960, May). *Master development plan report for Dulles International Airport.*
Ammann & Whitney, Eero Saarinen & Associates, Burns & McDonnell, & Ellery Husted. (1964). *Dulles International Airport: Master plan report.* Prepared for the Federal Aviation Agency. Washington, DC: Federal Aviation Agency.
Anderson, B. (2006). *Imagined communities: Reflections on the origin and spread of nationalism.* Rev. ed. London & New York: Verso.
Anonymous, & Thomas, A. R. (2001). *Air rage: Crisis in the skies.* New York, NY: Prometheus Books.
The architectural convention ends. (1907, October 8). *New York Times.*
Arend, G. (1979). *Air world's great airports: La Guardia 1939–1979.* Published in New York, NY, by Air Cargo News. Long Island, NY: LaGuardia Archives at LaGuardia Community College.
Arroyo, N. R., et al. (1953, May). Airport terminal buildings: Twentieth century building type. *Progressive Architecture.*
Authority–Leadership. (1971, April 11). *Tampa Tribune–Times.*

Bachman, L. R. (2002). *Integrated buildings: The systems basis of architecture.* New York City, NY: Wiley.

Backs LaGuardia as airport name. (1939, October 29). *New York Times.*

Bacon, R. F. [Director of the Bureau of National Capital Airports]. (1962a, March). [Letter to U.S. Congressman M. Gerhardt].

Bacon, R. F. [Director of the Bureau of National Capital Airports]. (1962b, April). [Letter to the Administrator of the FAA Memorandum No. 7941]. From the U.S. State Government.

Bacon, R. F. [Director of the Bureau of National Capital Airports]. (1962c, December). [Letter to the Administrator of the FAA].

Bacon, R. F. [Director of the Bureau of National Capital Airports]. (1963, May). [Letter to the Administrator of the FAA].

Bain, W. (Ed.). (2006). *The empire of security and the safety of the people.* Oxfordshire, UK: Taylor & Francis Group.

Ballard, J. G. (1997a, September). Airports. *The Observer*, p. 14.

Ballard, J. G. (1997b). Airports: The cities of the future. *Blueprint, 142.*

Bane, M. (1981). *The best li'l ole airport.* Tampa, FL: Hillsboro Printing and Engraving.

Bayor, R. H. (1993). *Fiorello La Guardia: Ethnicity and reform.* Arlington Heights, IL: Harlan Davidson.

Bednarek, J. R. D. (2001). *America's airports: Airfield development, 1918–1947.* College Station, TX: Texas A&M University Press.

Bednarek, J. R. D., & Bednarek, M. H. (2003). *Dreams of flight: General aviation in the United States.* College Station, TX: Texas A&M University Press.

Bennett, C. (2008). *The privacy advocates: Resisting the spread of surveillance.* Cambridge, MA: The MIT Press.

Bennett, C. J. (2008). *The privacy advocates: Resisting the spread of surveillance.* Cambridge, MA and London, UK: MIT Press.

Bennett, C. J., & Lyon, D. (Eds.). (2008). *Playing the identity card: Surveillance, security and identification in global perspective.* New York, NY: Routledge.

Bigo, D. (2004). Globalization of In-security: The field of professionals of unease management and the ban-opticon. *Traces: A Miltilingual Journal of Cultural Theory, 4*, 109–112.

Bigo, D., & Guild, E. (Eds.). (2005). *Controlling frontiers: Free movement into and within Europe.* Aldershot, Hants, UK and Burlington, VT: Ashgate.

Bigo, D., & Tsoukala, A. (2008). *Terror, insecurity and liberty: Illiberal practices of liberal regimes after 9/11.* New York, NY: Routledge.

Bijker, W. E. (1997). *On Bicycles, bakelites and bulbs: Toward a theory of socio-technical change.* Cambridge, MA and London, UK: MIT Press.

Bijker, W. E., Hughes, T. P., & Pinch, T. (Eds.). (1987). *The social construction of technological systems: New directions in the sociology and history of technology*. Cambridge, MA and London, UK: MIT Press.

Billington, D. (1983). *The tower and the bridge: The new art of structural art.* Princeton, NJ: Princeton University Press.

Bird, A. (2011). Thomas Kuhn and the concept of paradigm. In E. N. Zalta (Ed.), *The Stanford encyclopedia of philosophy*. Stanford, CA: Center for the Study of Language and Information.

Bliss, A. J. (2011). *Making a Sunbelt place: Tampa, Florida, 1923–1964* (Doctoral dissertation). Retrieved from Gradworks database (Document No. 3416656).

Block, J. V. (1968, December 2). *International airport planning as influenced by aircraft development.* London: Lecture to the Royal Aeronautical Society.

Block, J. V. (1971). *Airports and environment.* Edition Aéroports de Paris.

Blow, C. J. (1996). *Airport terminals* (2nd ed.). Oxford, UK: Butterworth–Heinemann.

Boyer, M. C. (2003). Aviation and the aerial view: Le Corbusier's spatial transformations in the 1930s and 1940s. *Diacritics, 33*(3–4), 99–116.

Boyer, P. (1985). *By the bomb's early light: American thought and culture at the dawn of the atomic age.* Chapel Hill, NC and London, UK: University of North Carolina Press.

Brancker, J. W. S. (1977). *IATA and what it does.* Leiden, Netherlands: A. W. Sijthoff.

Braniff Style. (1971, April 11). Tampa, you're beautiful [Advertisement]. *Tampa Tribune–Times.*

Braun, H.-J. (1995). From airships to airbus. In W. M. Leary (Ed.), *Airships to airbus: The history of civil and commercial aviation.* Washington, DC: Smithsonian Institution.

Brink Coridan, M. (2006, June). *Leigh Fisher associates: History and lore, as seen through the eyes of Marjorie Brink Coridan.* Publisher location and name unknown.

Brodherson, D. (1993). *What can't go up can't come down: The history of American airport policy, planning and design* (Doctoral dissertation). Ithaca, NY: Cornell University.

Brodherson, D. (1996). An airport in every city. In J. Zukowsky (Ed.), *Building for air travel: Architecture and design for commercial aviation.* Munich, Germany and New York, NY: Prestel and Chicago, IL: Art Institute of Chicago.

Brodsky, A. (2003). *The great mayor: Fiorello La Guardia and the making of the city of New York*. New York, NY: Truman Talley Books/St. Martin's Press.

Building types study: Planning of jet airports. (1960, March). *Architectural Record.*

Castells, M. (2000). *The rise of the network society*. Malden, MA: Blackwell.

Chadeau, E. (1995). History of airports. In W. M. Leary (Ed.), *Airships to airbus: The history of civil and commercial aviation*. Washington, DC: Smithsonian Institution.

Chandos, J. (1956). Passenger progress. In *London Airport: The official story of the new world air centre.* London, UK: Ministry of Transport and Civil Aviation.

Clack, V. A. (1998). *Airports.* Grove Art Online. City, ST: Oxford University Press.

Collins English Dictionary (6th rev. ed.). (2003).

Commission on Human Security of the Government of Japan. (2003, May). *Human security now: Protecting and empowering people.* Clash of civilizations.

Computer to speed airport baggage. (1970, March 15). *New York Times.*

Congratulations Tampa, the sky's your limit. (1971). Advertisement from TECO. *Tampa Tribune–Times.*

Construction drawing set of New York Municipal Airport and Seaport. (1939). *Avery drawings and archives*. Avery Library Columbia University.

Corn, J. J. (1983). *The winged gospel.* New York, NY: Oxford University Press.

Cortissoz, R. (1924). *Portrait of ten country houses by Delano & Aldrich. Drawings by Chester B. Price.* Cornell Library.

Courtwright, D. T. (2005). *Sky as frontier: Adventure, aviation and empire.* College Station, TX: Texas A&M University Press.

Crang, M. (2002). Between places: Producing hubs, flows, and networks. *Environment and Planning A, 34*(4), 569–574.

Crouch, T. D. (1981). *A dream of wings: Americans and the airplane 1875–1905.* New York, NY and London, UK: Norton.

Cullinan, W. E. (1962, December 10). *Memorandum: Report on first three-weeks operation.* Federal Aviation Agency.

Culver, F. R. (1996). *Planning and implementation for the Tampa airport system replacement program*. France: AFCET.

Curnow, A. (1997). Narita. In *Early days yet: New and collected poems 1941–1997* (p. 32). Manchester, UK: Carcanet.

Custom-built customs here. (1971, April 11). *Tampa Tribune–Times.*

The DC-10 will fit nicely at new terminal. (1971, April 11). *Tampa Tribune–Times.*

De Neufville, R. (1975). Airport systems planning. In R. Horonjeff (Ed.), *Planning and design of aiports* (2nd ed.). New York: McGraw-Hill.

Delano, W. (1939, November 5). [Letter to Col. Brehon B. Somervell]. *William Delano collection.* New York: Archives at Yale University Library.

Delius, P., & Slaski, J. (2005). *Airline design.* Dusseldorf, Germany and New York, NY: TeNeues.

Despite merger talk, Northeast moves ahead. (1971). *Tampa Tribune–Times.*

Doganis, R. (2004). *The airport business.* New York: Metropolitan Books.

Douglas, D. (1995). Airports as systems and systems of airports. In W. M. Leary (Ed.), *Airships to airbus: The history of civil and commercial aviation.* Washington, DC: Smithsonian Institution.

Douglas, D. G. (1996). *The invention of airports: A political, economic and technological history of airports in the United States: 1919–1939* (Doctoral dissertation). University of Pennsylvania, Philadelphia, PA.

Douglas, D. G. (2010, October). *Spectacle of system: Making of the modern airport.* Unpublished paper presented at Society for the History of Technology conference, Tacoma, WA.

Douglas, W. (1993, January 27). [Letter to Halaby Najeeb E.].

Downer, J. (2006). *The burden of proof: Regulating reliability in civil aviation.* (Doctoral dissertation). Cornell University, Ithaca, NY.

Downs, T. (2009, May 13). An introduction to airport design. *Dwell.*

Dulles National Airport makes hit with travelers. (1962, December). *St. Louis Globe Democrat.*

Dunbar, B. (1960, March). Airport breakthrough or bust… FAA's Dulles plan is different. *Airlift.*

Dunn, H. (1985). *Tampa: A pictorial history.* Norfolk, VA: Donning.

Edwards, B. (1998). *The modern terminal: New approaches to airport architecture* (Ch. 9). New York, NY: Routledge.

Edwards, B. (2005). *The modern airport terminal: New approaches to airport architecture* (2nd ed.). New York, NY: Spoon Press.

Eggebeen, J. (2007). *Airport age: Architecture and modernity in America* (Doctoral dissertation). City University of New York, New York, NY.

Evans, G. (2006). *Airport engineering: New Denver Airport, Colorado.* New York, NY: McGraw-Hill Encyclopedia of Science and Technology.

Those faraway places, strange-sounding names. (1971, April 11). *Tampa Tribune–Times.*

Federal Aviation Agency. (1961a). *Dulles International Airport: Fact sheet.*

Federal Aviation Agency. (1961b). *FAA's Office of Public Affairs Newsletter.* Washington, DC: Author.

Federal Aviation Agency. (1961c). The mobile lounge: Fact sheet. *FAA News.* Washington, DC: Office of Public Affairs.

Federal Aviation Agency [Executive Director]. (1962a). [Letter to the Officer of Public Affairs FAA].

Federal Aviation Agency. (1962b, November 17–18). *Dulles International Airport: Dedication program.*

Fellheimer & Wagner, Architects, & Odell, L.L. (1940, August). Projected airport. *Architectural Forum.*

First and Tampa has it. (1971, April 11). Advertisement from the members of Tampa's City Council. *Tampa Tribune–Times.*

Florida Photographic Collection, State Library & Archives of Florida.

Foucault, M. (1977). *Discipline and punish: The birth of the prison* (A. Sheridan, Trans.). New York, NY: Vintage Books, Random House.

Froesch, C., & Prokosch, W. (1946). *Airport planning.* New York, NY: Wiley.

Fuller, G., & Harley, R. (2004). Anatomy of an airport. In *Aviapolis: A book about airports.* London, UK: Black Dog.

The future! Thanks to men of vision. (1971). Advertisement from Tampa and Hillsborough County. *Tampa Tribune–Times.*

Gallison, P., & Roland, A. (2000). *Atmospheric flight in the twentieth century: Vol. 3. Archimedes new studies in the history and philosophy of science and technology.* New York, NY: Springer.

Garret, E. (1976). Designers put passengers first. *Tampa Tribune–Times.*

Garret, E. (1986, June 29). Tampa makes aviation history. *Tampa Tribune–Times.*

Garvey, W., & Fisher, D. (2002). *The age of flight.* Greensboro, NC: Pace.

Gateway to the sun, design of the future, a reality today. (1971, April 11). *Tampa Tribune–Times.*

Getting there's easy from Tampa, Delta to share airside no. 3. (1971, April 11). *Tampa Tribune–Times.*

Geraci, P. (1965, June). World's safest airport. *Rotarian.*

Gesar, A. (1994). *New York's airports: John F. Kennedy, Newark, LaGuardia.* New York: Pyramid Media Group.

Gilman, C. P. (1907, November 9). When we fly. *Harper's.*

Godwin Brown, K. (1994, June 6). World-famous TIA spawned from former WWII air base: Know your Tampa, a weekly history lesson. *North Tampa.*

Goodman, W. C. (2000). Transportation by air: Job growth moderates from stellar rates. *Aviation Employment.*

Gordon, A. (2004). *Naked airport.* New York, NY: Metropolitan Books.

Gorham, H. (1971a, April 11). Tampans dazzled by their airport. *Tampa Tribune.* New and Features, Section B.

Gorham, H. (1971b, April 11). They came, they saw, they liked. *Tampa Tribune–Times.*

Gorham, H. (1971c, April 12). New airport jammed on open house: They came in greater thousands. *Tampa Tribune.*

Gottdiener, M. (2001). *Life in the air: Surviving the new culture of air travel.* New York, NY: Rowman & Littlefield.

Graff, C. (2008). *Boeing field.* Charleston, SC: Arcadia.

Graham, A. (2001). *Managing airports: An international perspective.* Oxford, UK and Boston, MA: Butterworth–Heinemann.

The greening of an airport. (1971). *Tampa Tribune–Times.*

Growth in the air. (1976, November 11). *Tampa Tribune.*

Hailey, A. (1968). *Airport.* Mishawaka, IN: Doubleday & Company.

Halaby, N. E. (1962a, January 4). *Letter to Morris Strassman.*

Halaby, N. E. (1962b, March 2). *Letter to Mr. Gerhardt.*

Halaby, N. E. (1962c, November). Saluting Washington's new jetport. *Airplanes.*

Halaby, N. E. (1963, February 21). *Letter to William Douglas.*

Hall, R. (2015). *The transparent traveler.* Durham, NC and London, UK: Duke University Press.

Hard work and patience. (1971, April 11). *Tampa Tribune–Times.*

Harmel, K., & Spadanuta, L. (2006, September 9). Mickey wants your prints. *Globe and Mail,* p. T9. Toronto.

Harris, A., & Friedlander, L. (Eds.). (2004). *Arrivals and departures: The airport pictures of Garry Winogrand.* New York, NY: Distributed Art.

Hasan, D. (2010, May 25). Airport design vocabulary that moves you emotionally, today and tomorrow. *Unique travel destinations.* Retrieved from http://uniquetraveldestinations.wordpress.com/2010/05/25/airport-design-vocabulary-that-moves-you-emotionally-today-and-tomorrow-2/.

Haugland, V. K. (1962, December). The people likes it: New Dulles Airport starts out as $110 million hit. *New Haven Register.*

Hawes, L. (1978). Drew Field was a world unto itself. *Tampa Tribune.*

Heidegger, M. (1982). *The question concerning technology.* New York, NY: HarperCollins.

Higham, R. (1995). A matter of utmost urgency. In W. M. Leary (Ed.), *Airships to airbus: The history of civil and commercial aviation.* Washington, DC: Smithsonian Institution.

Hill, J. (1996, April 14). Tampa airport looks good for its years. *Tampa Tribune.*

Hillsborough County Aviation Authority. (1965). *A new approach to jet age air terminal development: Progress report 1961–65.* Tampa, FL: Author.

History made in 1914: More than 57 years ago, Tampa Bay had air service. (1971, April 11). *Tampa Tribune–Times.*

Hobbs, G. W. (1963, May 14). *Memorandum: Airline practice regarding through passengers at Dulles International Airport.* Federal Aviation Agency.

Holden, H. M. (2009). *Newark airport.* Charleston, SC: Arcadia.

Homan, L. M., & Reilly, T. (2000). *Pan Am.* Charleston, SC: Arcadia.

Van Houtum, H., Kramsch, O., & Zierhofer, W. (Eds.). (2005). *B/ordering space.* Border regions series. Burlington, VT: Ashgate.

Van Houtum, H., & van de Velde, M. (2004). The power of cross-border labour immobility. *Tijdschrift voor Economische en Sociale Geografie (TESG), 95*(1), 100–107.

Van Houtum, H., & van Naerssen, T. (2002). Bordering, ordering and othering. *Tijdschrift Voor Economische En Sociale Geografie, 93*(2), 129.

Huettel, S. (1995, September 9). Time keeps on moving: Tampa International Airport will retire some of the first shuttles ever installed at an airport this week. *Tampa Tribune.*

Hughes, T. P. (1983a). *The evolution of large technological systems: The social construction of technological systems.* Cambridge, MA: MIT Press.

Hughes, T. P. (1983b). *Networks of power: Electrification in western society 1880–1930.* Baltimore, MD and London, UK: Johns Hopkins University Press.

Hughes, T. P. (1987). *The evolution of large technological systems: The social construction of technological systems.* Cambridge, MA: MIT Press.

If you're flying to Washington. (1962). *U.S. News & World Report.*

If you're flying to Washington: What to expect at world's biggest airport. (1963). *U.S. News & World Report.*

Iglesia, R. E. J. (1966). *Eero Saarinen.* Buenos Aires, Argentina: Instituto de Arte Americano e Investigaciones Estéticas.

An impression: The city will have to keep up with its airport. (1971, April 11). *Tampa Tribune–Times.*

International Airport, Tampa, Florida. (2002, January 18). *La Gaceta.*

International Airport Commemorative Edition. (1971, April 11). *Tampa Tribune–Times.*

International Civil Aviation Organization. (1953). *Aerodromes: Annex 14 to the Convention on International Civil Aviation* (2nd ed.). International Standard and Recommended Practices. Montreal, QC: Author.

International Civil Aviation Organization. (1974). *Annex 17 to the Convention on International Civil Aviation* (1st ed.). International Standard and Recommended Practices. Montreal, QC: Author.

International Civil Aviation Organization. (2005). *ICAO's accident prevention program manual* (Document No. 9422). Section 5:3.

IRBMW. (2011, April). *6 years old girl physical inspection (U.S. Airport).* Retrieved January 15, 2013, from http://www.youtube.com/watch?v=ApmBa-ZSh9E.

J. D. Power and Associates. (2010). *2010 North America airport consumer satisfaction study.* Retrieved from http://www.jdpower.com/content/study/7ldlX-0O/2010-north-america-airport-satisfaction-study.htm?page=3.

Jet-age progress. (1959a, April 12). *New York Post.*

Jet-age airport. (1959b, December). Bonanza from Billy. Art. *New Yorker.*

A jet airport opens. (1962, November 26–30). *My Weekly Reader: The Children's Newspaper.*

Johnson, P. (1946). Postwar planning reconversion. *Aviation Digest index.* Washington, DC: U.S. National Research Bureau (Journal).

Jones, G. (2003). *Delta air lines: 75 years of airline excellence.* Charleston, SC: Arcadia.

Karsner, D. (1993). *"Leaving on a Jet Plane": Commercial aviation, airports and post-industrial American society, 1933–1970* (Doctoral dissertation). Temple University, Philadelphia, PA.

Kennedy, D. M. (2005). *Freedom from fear: The American people in depression and war 1929–1945.* New York, NY and Oxford, UK: Oxford University Press.

Kessner, T. (1989). *Fiorello H. La Guardia and the making of the modern New York.* New York, NY: McGraw-Hill.

Khurana, K. C. (2009). *Aviation management, global perspectives.* New Delhi, India: Global India Publications.

Kinney, J. R. (2008). *Airplanes: The life story of a technology.* Baltimore, MD: Johns Hopkins University Press.

Kline, R., & Pinch, T. (1996). Users as agents of technological change: The social construction of the automobile in the rural United States. *Technology and Culture, 37*(4), 763–795.

Lackney, J. A. (1999, September). *Assessing school facilities for learning/assessing the impact of the physical environment on the educational process.* Mississippi State, Miss.: Educational Design Institute.

Lackney, J. A., & Tanner, C. K. (1999). A history of the studio based learning model. In C. K. Tanner & J. A. Lackney (Eds.), *Educational facilities planning.* Boston, MA: Allyn & Bacon.

Lahav, G. (2008). Mobility and border security: The U.S. aviation system, the state, and the rise of public-private partnerships. In M. B. Salter (Ed.), *Politics at the airport.* Minneapolis, MN and London: University of Minnesota Press.

Landside, airside (Def.1). (2003). In *Collins English dictionary* (6th rev. ed.).

Latour, B. (1986). *Laboratory Life*. Princeton, NJ: Princeton University Press.

Latour, B. (1993). *La Clef de Berlin et Autres Leecons d'un Amateur de Sciences*. Paris: La Decouverte.

Latour, B. (1996). *Aramis, or, the love of technology*. Cambridge, MA: Harvard University Press.

Launius, R. D., & Bednarek, J. R. D. (2003). *Reconsidering a century of flight*. Chapel Hill, NC and London, UK: University of North Carolina Press.

Leary, W. M. (Ed.). (1995). *From airships to airbus: The history of civil and commercial aviation: Vol. 1. Infrastructure and environment*. Washington, DC: Smithsonian Institution.

L' ECLAT rédaction. (2011, February 13). *Charles de Gaulle Terminal 1: l'aérogare mal-aimée*. L'ECLAT Regards du Actualité.

Leggett, B. (2012). Departures: The novel, the non-place and the airport. *Alluvium, 1*(4), no page.

Leigh Fisher Associates, Inc. (1963a, April). *Study of U.S. airport terminal building as a basis for planning new passenger terminal facilities at Tampa International Airport* (Vol. 1). San Francisco, CA: Author.

Leigh Fisher Associates, Inc. (1963b, July). *Recommended planning criteria for new passenger terminal facilities at Tampa International Airport* (Vol. 2). San Francisco, CA: Author.

Levine, M. E. (2008). *Airport congestion: When theory meets reality*. NELLCO Legal Scholarship Repository, New York University.

Ley, D., & Cybriwsky, R. (1974). Urban graffiti as territorial markers. *Annals of the Association of American Geographers, 64*(4), 491–505.

The lights go on at Dulles Airport. (1962, September 14). *Washington Post*.

van der Linden, F. R. (2002). *Airlines and air mail: The post office and the birth of the commercial aviation industry*. Lexington, KY: University Press of Kentucky.

Lloyd, J. (2002). Departing sovereignty. *Borderlines e-journal, 1*(2), 1–12.

Lloyd, J. (2003). Dwelltime: Airport technology, travel, and consumption. *Space and Culture, 6*(2), 93–109.

Loading made easier. (1952). *Flight*.

Löfgren, O. (1999). Crossing borders: The nationalization of anxiety. *Ethnologia Scandinavica, 29*, 5–27.

Logan still a winner in time. (1965). *Boston Globe*.

A look at the newest thing in airports. (1962, September 10). *U.S. News & World Report*.

Lyon, D. (2007). Surveillances studies: An overview. Cambridge, UK and Malden, MA: Polity Press.

Lyon, D. (2008). Filtering flows, friends, and foes: Global surveillance. In M. B. Salter (Ed.), *Politics at the airport* (Ch. 2). Minneapolis, MN and London: University of Minnesota Press.

Manners, W. (1976). *Patience and fortitude: Fiorello LaGuardia*. New York, NY and London, UK: Harcourt, Brace & Jovanovich.

Marquez, V. (2007, November). *Why new materials take old shapes: The case of the first iron-bridge in 1779*. Conference paper presented at the 2007 meeting of the Swiss Association for the Study of Science, Zurich.

McCarthy, K. (2003). *Aviation in Florida*. Sarasota, FL: Pineapple Press.

McCulloch, A. (2008, September). The world's coolest airport terminals. *Forbes Traveler*.

McDevitt & Street Company. (1971). We are proud to be part of the new Tampa International Airport [Advertisement]. *Tampa Tribune–Times*.

Meyer, J., Kain, J. F., & Wohl, M. (1965). *Technology and urban transportation*. Cambridge, MA: Harvard University Press.

Mobile lounge-ramp ends walk to planes. (1962). *Birmingham eccentric*.

Mormino, G. R., & Pizzo, A. P. (1983). *Tampa*. Tampa Bay, FL: Continental Heritage.

Mumford, L. (1934). *Technics & civilization*. Chicago, IL: University of Chicago Press.

Mumford, L. (1939, December). The sky line. *The New Yorker*, p. 49.

National Commission on Terrorist Attacks Upon the United States. (2004). *9/11 and terrorist travel: A staff report of the National Commission on Terrorist Attacks Upon the United States*. Franklin, TN: Hillsboro.

Nelkin, D. (1974). *Jetport: The Boston airport controversy*. New Brunswick, NJ: Transaction Books.

New 180-foot control tower due, to cost $1.4 million. (1971, April 11). *Tampa Tribune–Times*.

New airport serves Pinellas County too. (1971, April 11). *Tampa Tribune–Times*.

New terminal no overnight miracle. (1971, April 11). *Tampa Tribune–Times*.

New thinking of airport terminals. (1956, June). *Architectural Forum*.

The New York City Landmarks Preservation Commission. (1980). *Report*.

New York Municipal Airport. (1938). *A detailed description of the great air terminal by the Work Projects Administration* [Original document]. I.T.S. Library, University of California at Berkeley.

No Longer Possible. (1995, February 15). Advertisement from Tampa International Airport. *Tampa Tribune–Times*.

Nordle, P., National Aviation Facilities Experimental Center, Human Sciences Research, Inc., et al. (1961). *Airport transportation: A study of transportation means between airports and the metropolitan areas they serve.* Arlington, VA: Human Sciences Research.

North America Airport Consumer Satisfaction Study. (2010). J. D. Power and Associates. Retrieved from http://www.jdpower.com/content/study/7ldlX-0O/2010-north-america-airport-satisfaction-study.htm?page=3.

Not just a hotel. (1971, April 11). *Tampa Tribune–Times.*

Nye, D. (1994). *American technological sublime.* Cambridge, MA: MIT Press.

In October 1971, Eastern Airlines will fly to another world. (1971). Advertisement from Eastern Airlines. *Tampa Tribune–Times.*

Odell, L. L. (1939, October 1). Letter to William Delano. *William Delano collection.* Archives at Yale University Library.

O'Donnell, B. (1991, May 14). Designs readied for TIA airside. *Tribune.*

Onnée, J.-F. (2004, December). *Paris—Charles de Gaulle Airport: A critique of the development of France's first airport.* Term paper for MIT Airport Systems Planning, Design and Management.

Our part of the finest Tampa International Airport. (1971). Advertisement from J. A. Jones. *Tampa Tribune–Times.*

Paasi, A. (1996). *Territories, boundaries and consciousnesses: The changing geographies of the Finnish-Russian boundary.* New York, NY: Wiley.

Pascoe, D. (2001). *Airspaces.* London: Reaktion Books.

Pasquier, R. D. (1939). [Letter to the Office Central des Chemins de Fer Federaux in Berne, Switzerland]. *William Delano collection.* Archives at Yale University Library.

Passenger walking distances, Kodak color patch. (1959). *Saarinen Archive.* Manuscripts and Archives at Yale University Library.

Passi, Anssi. (1996). *Territories, boundaries and conciousness: The changing geographies of the finish-russian boundary.* New York, NY: J. Wiley and Sons.

The passengers love Boeing's big 747, but it has faults. (1971, April 11). *Tampa Tribune–Times.*

Peck, M. C. (2005). *Washington Dulles International Airport.* Charleston, SC: Arcadia.

Pellow, D. (Ed.). (1996). *Setting boundaries: The anthropology of spatial and social organization.* Westport, CT and London, UK: Bergin & Garvey.

Pennoyer, P., & Walker, A. (2003). *The architecture of Delano & Aldrich.* New York, NY and London, UK: Norton.

Perkins, C. (1907). When we fly: How the accomplishment of aerial navigation will make necessary a revision of human laws and customs. *Haper's*, pp. 2–3.

Perlman, H. (2004). *Airports: A century of architecture*. London, UK: Lawrence King.

Pescador, K., Aldrich, M., & San Diego Air and Space Museum. (2008). *Consolidated aircraft corporation*. Charleston, SC: Arcadia.

Peterson, S. (1996, January 2). Checkpoint Charlies dash British convoy's blithe spirits. *Christian Science Monitor*, p. 4.

Pinch, T., & Bijker, W. (1987). The social construction of facts and artifacts: Or how the sociology of science and technology might benefit each other. In W. E. Bijker, T. P. Hughes, & T. J. Pinch (Eds.), *The social construction of technological systems*. Cambridge, MA: MIT Press.

Planning of jet airports, passenger terminal building design principles. (1960, March). *Architectural Record*.

Poh, E. [Assistant Director of ICAO Affairs]. (2007, April 9–13). *Lecture on Strategic Airport Management Programme, Airport Planning and Terminal Design*.

Pommer, R. (1983). Revising modernist history: The architecture of the 1920s and 1930s. *Art Journal, 43*(2), 158–169.

Port Tampa City will fire mayor if he won't quit. (1930, February 14). *Tampa Morning Tribune*.

Progress is a new air terminal. (1971, April 11). *Tampa Tribune–Times*.

Projected Airport by Fellheimer & Wagner Architects, L.L. Odell Aviation Consultant. (1940, August). *Architectural Forum*, pp. 85–87.

Prokosch, W. (1951, January). *Airport design: Its architectural aspects* (Architectural Record's Building Types Study, Number 170). Architectural Record.

Proust, M. (1954). *Swann's Way*. New York. Heritage Press.

Proust, M. (1992). *In search of lost time* (D. J. Enright, Rev., Vol. 7). London: Chatto and Windus and New York: Modern Library.

Quarter, T. (1962). Now there are wheels under the airport lounge. *Steel Horizons*.

Rane, J. (2011, November 17). Web poll results: 10 of the world's most loved airports. *CNN Travel*. Retrieved from http://travel.cnn.com/explorations/life/10-most-loved-airports-981939.

Reitman, I., & Reitman, J. (2009). *Up in the air [motion picture]*. USA: Dream Work Pictures.

Riley, A. A. (1962, November 25). At last—An airport that's easy on the feet. *Science in Industry*.

Rodrigue, J.-P., Comtois, C., & Slack, B. (2009). *The geography of transport systems* (2nd ed.). New York, NY: Routledge.

Rogers, E. (1995). *Diffusion of Innovations*. New York, NY: The Free Press.

Rogers, E. M. (1962). *Diffusion of innovations* (4th ed.). New York, NY: Free Press.

Roman, A. (2003). *Eero Saarinen: An architecture of multiplicity.* New York, NY: Princeton Architectural Press.

Rosenberg, N. (1982). *Inside the black box: Technology and Economics.* Cambridge, UK and New York: Cambridge University Press.

Rosenberg, N. (1994). *Exploring the black box: Technology, economics and history.* Cambridge, UK and New York: Cambridge University Press.

Rowley, J., & Slack, F. (1999). The retail experience in airport departure lounges: Reaching for timelessness and placelessness. *International Marketing Review, 16*(4/5), 363–375.

Ruark, R. C. (1961, May 2). The airport dilemma. *New York World Telegram.*

Eero Saarinen Collection. (1938–1962). *Printouts of advertisements by Chrysler, publicizing the virtues of the Mobile Lounge (1961–1963).* Manuscripts and Archives, Yale University Library.

Saarinen, E. (1962, November). Architect of Dulles Terminal. *Airplanes.*

Saarinen, E. 3080–3133. Box 463, Doc. 59. Collection Manuscripts & Archives, Yale University.

Salter, M. (2003). *Rights of passage: The passport in international relationships.* London, UK: Lynne Rienner Publishers.

Salter, M. B. (2003). *Rights of passage: The passport in international relations.* Boulder, CO: Lynne Rienner.

Salter, M. B. (2005). At the threshold of security: A theory of international borders. In E. Zureik & M. Salter (Eds.), *Global surveillance and policing: Borders, identity, security.* Willan: Cullompton, UK.

Salter, M. B. (2006). The global visa regime and the political technologies of the international self: Borders, bodies, biopolitics. *Alternatives, 31*(2), 167–189.

Salter, M. B. (2007). Governmentalities of an airport: Heterotopia and confession. *International Political Sociology, 1*(1), 46–67. Reprinted in M. B Salter (Ed.). (2008). Introduction: Airport assemblage. *Politics at the airport.* Minneapolis, MN and London: University of Minnesota.

Salter, M. B. (2008a). The global airport: Managing space, speed, and security. In M. B. Salter (Ed.), *Politics at the airport.* Minneapolis, MN and London: University of Minnesota Press.

Salter, M. B. (Ed.). (2008b). *Politics at the airport.* Minneapolis, MN and London, UK: University of Minnesota Press.

A Santa in April, officially speaking, smiles are in order. (1971, April 11). *Tampa Tribune–Times.*

Sayles, J. (Director). (2009). *Limbo, a condition of unknowable outcome* [Film]. New York, NY: Green/Renzi.

Schaberg, C. (2017). *Airportness, the nature of flight.* New York, NY: Bloomsbury Academic.

Scott, J. C. (1998). *Seeing like a state: How certain schemes to improve the human condition have failed.* New Haven, CT and London, UK: Yale University Press.

Scrapbooks 54–55, Wagner–LaGuardia Archives at LaGuardia Community College, Long Island City, NY.

Scullin, G. (1968). *International airport: The story of Kennedy Airport and U.S. Commercial Aviation.* Boston, MA and Toronto, ON: Little, Brown.

Shaffer, J. H. (1970a, June 25). [Letter to Senator William B. Spong, Jr.].

Shaffer, J. M. (1970b, March 7). [Letter to the Secretary of Transportation, for proposed baggage handling system, Seattle-Tacoma International Airport].

Sharp, D. (1972). Decade 1950. In D. Sharp (Ed.), *Twentieth-century architecture: A visual history* (pp. 175–232). Sydney, Australia: Images.

Shuttle comes in for a landing. (1987, June 10). *Tribune.*

Simms, M. (2007, October/November). Perfect attendants. *Bust.*

Sismondo, S. (2009). *An introduction to science and technology studies.* Malden, MA: Blackwell.

Smith, Hinchman & Grylls, Inc., Architects and Engineers. (1947, October). Establishing an airport planning program. *Architectural Record,* pp. 90–94.

Smith, P., & Toulier, B. (2000). *Berlin Tempelhof, Liverpool Speke, Paris Le Bourget: Casisse national des monuments historiques et des sites 200.* Paris, France: Ed. du Patrimoine.

Stanton, J., Chango, M., & Owens, J. (2008). *ICAO and the biometric RFID passport: History and analysis.* Syracuse University, School of Information Studies.

Stapleton, D. H. (1987). *The transfers of early industrial technologies to America.* Philadelphia, PA: American Philosophical Society.

Star, S., & Griesemer, J. (1989). Institutional ecology, "translations" and boundary objects: Amateurs and professionals in Berkeley's Museum of Vertebrate Zoology, 1907–39. *Social Studies of Science, 19*(3), 387–420.

Starting today… through the New Tampa International Airport. (1971, April 11). Advertisement from C. A. Fielland, Inc. *Tampa Tribune–Times.*

State's skies getting congested as volume of air traffic grows. (1989, June 5). *Florida Times–Union.*

Stoff, J. (2004). *Long Island airports*. Charleston, SC: Arcadia.
Stoff, J. (2008). *LaGuardia airport*. Charleston, SC: Arcadia.
Stover, D. (2003, March). Airport of the future: Science Newsfront. *Popular Science*.
Sutton, H. (1962a). Mobile lounge replaces cardiac causeway. *Travel Notes*.
Sutton, H. (1962b). New job: Bustronaut. *Travel Notes*.
Swatek, P. M. (1962, March 23). [Letter on national airport problems]. National Archives.
Szurovy, G. (2002). *The art of airways*. St. Paul, MN: MBI.
Szurovy, G. (2003). *The American airport*. St. Paul, MN: MBI.
Tampa Airport Marriot Hotel. (1983, March 27). *Tribune*.
Tampa International Airport. (1960, November 21). *Tribune*.
Tampa International Airport. (1972). [Brochure on Tampa parking system and terminal].
Tampa International Airport—One of the world's best, a destination, a gateway. (1976, February 19). *Tampa Tribune*.
Tampa's jetset. (1971, April 11). *Tampa Tribune–Times*.
Tampa's new crown jewel. (1971, April 11). *Tampa Tribune–Times*.
Taylor, E. (1986, June 29). Airport wins praise from airlines. *Tampa Tribune–Times*.
Taylor, E. (1991, June 23). TIA airport operations still flying high. *Tampa Tribune–Times*.
Technology and culture. (2008). *International Quarterly of the Society for the History of Technology, 49*(4).
Temko, A. (1962). *Eero Saarinen*. New York, NY: George Braziller.
Terminal sprawl, no ailment here. (1971, April 11). *Tampa Tribune–Times*.
Thibault, P. M. (2003). *Le roman d'Air France*. Paris, France: IME Gallimard.
This is how you board a plane at new Dulles Airport. (1962, November 11). *Sunday Star*. Washington, DC.
Thompson, C. (1962, September 1). *The AOPA airport letter.* Washington, DC: Aircraft Owners & Pilots Association.
Time study—Mobile lounge, Kodak color patch. (1959). *Saarinen Archive*. Manuscripts and Archives, Yale University Library.
To simulate area growth, just build a new airport. (1971, April 11). *Tampa Tribune–Times*.
Toulier, B. (2000). *Berlin Tempelhof, Liverpool Speke, Paris Le Bourget: Casisse national des monuments historiques et des sites*. Paris: Patrimoine.
Turkle, S. (1995). *Life on the screen*. New York, NY: Touchstone, Simon & Schuster.

TWA introduced jets to Tampa. (1971, April 11). *Tampa Tribune–Times.*

Two-county controversy fades as both prosper. (1971, April 11). *Tampa Tribune–Times.*

Unique mobile lounges: Dulles National Airport makes hit with travelers. (1962, December 10). *St. Louis Globe-Democrat.*

United: A Tampa mainstay since 1961. (1971, April 11). *Tampa Tribune–Times.*

Urwin, G. J. W. (2002). *Facing fearful odds: The siege of Wake Island.* Lincoln, NE: University of Nebraska Press.

U.S. Civil Aeronautics Administration. (1953, April). *Airport terminal buildings.* Washington, DC: U.S. Department of Commerce.

The U.S. President's Airport Commission. (1958). *The airport and its neighbors.* Report of the U.S. President's Airport Commission. Washington, DC: Government Printing Office.

Van Houtum, H., & van Naerssen, A. L. (2002). *Bordering, ordering and othering.* Amsterdam, NL: Royal Dutch Geographic Society.

Vincenti, W. G. (1976). *What engineers know and how they know it: Analytical studies of aeronautical history.* Baltimore, MD and London, UK: Johns Hopkins University Press.

Voigt, W. (1996). From hippodromes to aerodromes. In J. Zukowsky (Ed.), *Building for air travel: Architecture and design for commercial aviation.* Munich, Germany, New York, NY: Prestel and Chicago, IL: Art Institute of Chicago.

Waiting room on wheels. (1961, December). *Contract.*

Waresh, J. (1987, April 20). Bean helped Tampa airport win its wings. *Tampa Tribune.*

Washington National Airport. (1941, October). *Architectural Record, 90*(4).

Weber, R. (1997). Manufacturing gender in commercial and military cockpit design in science. *Technology & Human Values, 22*(2), 235–253.

Westinghouse Company. (1971). [Illustrated advertisement insert]. *Times.* Tampa.

Wilson, T. M., & Donnan, H. (Eds.). (1998). *Border identities: Nation and state at international frontiers.* Cambridge, UK: Cambridge University Press.

Winner, L. (1980). Do artifacts have politics? *Daedalus, 109*(1), 121–136. http://www.jstor.org/stable/20024652.

Winner, L. (1986). Do artifacts have politics? In L. Winner (Ed.), *The whale and the reactor: A search for limits in an age of high technology.* Chicago, IL: University of Chicago Press.

Wittgenstein, L. (1970). *Zettel.* Berkeley, CA: University of California Press.

Wohl, R. (2005). *The spectacle of flight.* New Haven, CT and London, UK: Yale University Press.

Work Projects Administration. (1938). *A detailed description of the great air terminal.* New York Municipal Airport. I.T.S. Library University of California at Berkeley.

Work Projects Administration. (1939a, July 1). *Report of activities.* New York Municipal Airport. I.T.S. Library University of California at Berkeley.

Work Projects Administration. (1939b, October 5). Dedication. *Program of events.* New York Municipal Airport. I.T.S. Library University of California at Berkeley.

Yen, A. M., Henderson, C., Sakasita, M., Roddin, M., Cronin, R., Ellis, H., U.S. Urban Mass Transportation Administration, & SRI International. (1977, December). *Assessment of the passenger shuttle system (PSS) at Tampa International Airport.* Washington, DC: Urban Mass Transportation Administration, and Springfield, VA. Available through the National Technical Information Service.

You ride to plane in mobile lounge. (1962, November). *Sunday Star.* Washington, DC.

Zukowsky, J. (Ed.). (1996). *Building for air travel: Architecture and design for commercial aviation.* Munich, Germany and New York, NY: Prestel and Chicago, IL: Art Institute of Chicago.

Index

A

Abramowitz 92
Abu Dhabi Airport 200
actors 10, 16, 25, 82, 83, 102, 108, 114, 115, 125, 127–129, 156, 165, 168–170, 197, 210, 212, 248
actorship 182
aerodrome 10, 26, 28, 35, 48
Aéroports de Paris (ADP) 16, 183, 188–190
agency 49, 84, 100, 101, 112, 115, 177
agents 31, 175, 177, 210
air 2, 8, 11, 25, 27–33, 35–40, 45, 46, 49, 52, 54, 57–60, 66, 74–76, 79, 81, 83, 85–88, 91–96, 100, 101, 108, 110, 111, 118, 119, 127, 128, 130, 134, 135, 137, 139, 141, 145, 148, 161, 164, 180, 181, 186, 187, 189, 190, 200, 208, 214, 215, 222, 223, 230, 233, 234, 238, 248
air bridge 156, 198
aircraft 1, 3, 6, 8–11, 14, 15, 18, 22, 24–28, 30, 31, 33–35, 37, 41, 46, 52–57, 61, 63, 73, 74, 76, 81, 82, 84, 90–92, 94, 95, 97, 98, 101, 104, 107, 110–114, 116, 117, 119, 120, 127, 128, 131, 135–137, 140, 141, 146–148, 150, 151, 153, 157, 159, 161–164, 170, 178, 182–185, 188, 189, 192, 195, 198, 206, 208, 209, 215, 220, 229, 233, 246, 248, 253
airdromes 32
airfield 8, 25, 27, 28, 39, 45, 46, 57, 63, 64, 98, 100, 133, 144, 150, 155, 156, 159, 183
airmail 15, 34–38, 66

V. Marquez, *Landside | Airside*, https://doi.org/10.1007/978-981-13-3362-0

airmindedness 10, 34, 37
airplane 8, 15, 21, 26–31, 34, 37, 45, 46, 53, 54, 76, 79, 95, 98, 100, 110, 135, 152, 155, 156
airport 1–18, 21–23, 26, 27, 31–45, 47–49, 51–66, 71–87, 89–108, 110, 112–121, 125–149, 151–157, 159–172, 175–189, 191–197, 199–201, 203–240, 245–251, 253–256
airport planning 5, 7, 12, 16, 17, 23, 40, 43, 46, 48, 54, 62, 71, 73, 76–80, 85, 91, 93, 99, 106, 125–129, 133, 144, 152, 157, 159, 162, 163, 171, 176, 179, 181, 185, 187, 189, 193, 206, 246, 248, 255
airside 3–8, 10–13, 16–18, 22, 27–29, 31, 39, 48, 60, 63, 65, 66, 72–75, 77, 80–82, 84–92, 94–97, 99, 102, 104, 106, 108, 112–114, 116–121, 125, 127–129, 131, 132, 134–137, 139, 141, 143, 144, 146–157, 160, 161, 165–171, 175, 176, 178–182, 184–188, 190–198, 200, 201, 203–207, 211–213, 215–218, 220, 221, 223, 224, 226, 228–239, 245–251, 253–255
Akrich, Madelaine 17, 96, 108, 254
Amsterdam Airport 211
Anchorage Airport 104
Annex14 85–89, 147, 188, 215
Annex17 120, 207, 213, 215
Antalya Airport 207
apron 8, 12, 23, 25, 28, 29, 45, 50–56, 86, 88, 90, 92, 94, 95, 109, 111, 112, 129, 146, 147, 151, 159, 164, 178, 183, 187, 196, 198, 229
architect 4, 11, 22, 23, 42, 44, 47, 48, 52, 53, 58, 66, 74, 80, 82, 92, 102–104, 106, 132, 183, 187, 189, 194, 195, 208
Atlanta Airport 133, 154, 171, 179, 191, 238
author 34, 40, 52, 59, 66, 74, 81, 98, 134, 148, 187, 208, 211, 215, 223–225, 230
autodrome 28
automatic 48, 61, 62, 73, 184, 237
aviation 2, 8, 14, 15, 18, 21, 23–28, 30–32, 34–37, 40, 41, 43, 46, 47, 51, 52, 56, 59, 61, 62, 64, 71, 75, 77, 78, 85, 86, 91–93, 95, 99, 116, 119, 120, 130, 132–135, 141, 142, 145, 146, 164, 165, 178, 198, 206, 209, 213, 215, 249, 253

B

baggage 47, 48, 52–54, 56, 61, 82–84, 88–90, 94, 110, 112, 114, 131, 140, 143, 151, 162, 195, 198, 214, 220, 253
Bangkok Airport 212
Ben Gurion Airport 194, 198, 206, 210, 238
Bijker, Wiebe 18, 107, 108, 127, 160, 165, 168, 247, 248, 255
biometrics 211, 213, 239
black boxes 7, 15, 16, 216, 217
Bogota Airport 238
border 13, 156, 200, 207, 211, 218–221, 223, 224, 232, 233, 236, 250

borderless 219
borderline 10, 14, 35, 57, 66, 95, 229, 230, 236
boundary 2–8, 10–14, 16–18, 22, 27–29, 48, 61, 65, 72–75, 80, 82, 84, 87–92, 95–97, 99, 102, 106, 112–114, 116–121, 125, 127, 128, 131, 150, 151, 153, 155–157, 165, 166, 168, 175, 180–182, 185–187, 191, 192, 194–201, 203–208, 212, 213, 215–220, 223, 225, 229, 230, 232, 234–236, 238, 239, 245–251, 254, 256
boundary object 3, 16, 254
bridge 96, 100, 101, 118, 141, 150, 156, 162
building 9, 11, 22, 26, 31, 40, 43, 45, 47, 50, 53, 54, 58–64, 73, 74, 76, 82, 86–88, 90–93, 95–98, 105–107, 110, 112–116, 118–120, 128, 129, 134, 137, 138, 140, 146–148, 152, 153, 156, 160–163, 176, 178, 182–184, 188–190, 193–196, 198, 221, 239
Burbank Airport 104

C

Cairo Airport 200
cargo 54, 79, 82, 96, 147, 148
cathedrals 11, 74, 238
change 3, 6, 11, 15, 17, 31, 34, 56, 73, 75, 77, 94, 95, 98, 108, 114, 117, 128, 140, 146, 152, 153, 161, 165, 177, 180, 185, 191, 206, 209, 211, 216, 230, 249, 253, 255
Changi Airport 211
channel 50, 94, 210, 226
Charleston Airport 104
check-in 54, 83, 89, 93, 100, 112, 162, 189, 195, 218, 226, 227
Chicago Airport 92, 96, 100, 104, 105, 126, 138, 143, 183, 215
Chicago Municipal Airport 92
city 9–11, 25, 26, 30, 31, 37–39, 42, 47, 48, 52, 53, 55, 57, 58, 64, 74, 78, 79, 92, 99, 129, 143, 177, 179, 182, 183, 187, 189, 194–196, 198, 214, 230
Civil Aviation Authority 215
civilian 75, 93, 206
clean 30, 239
clear 5, 28, 30, 32, 52, 61, 63, 79, 84, 87–89, 92, 94, 97, 113, 130, 132, 154, 161, 163, 168, 197, 209, 215, 226–228, 231, 234
commercial aviation 11, 27, 31, 32, 37, 64, 74, 75
concept 3, 4, 7, 12, 15, 17, 22, 28, 31, 34, 61, 75, 77–79, 89, 91, 92, 94, 102, 106–108, 110, 112, 113, 116–118, 120, 125, 127, 128, 130–133, 135, 136, 139–141, 143, 144, 146, 148, 150, 151, 156, 157, 159, 161, 164, 166–169, 171, 177, 190, 193, 216–218, 226, 239, 248, 249, 254, 255
concessions 76, 81, 163, 164
concourse 51, 58, 61, 87, 90, 91, 113, 114, 128, 148, 151, 154, 193, 195, 216, 227
congestion 2, 12, 55, 71, 73, 93, 99, 128, 130, 136, 144, 145, 161,

164, 166, 176, 180–182, 185, 191, 197, 200, 210, 227, 247, 248, 250, 254
containment 5, 6, 13, 89, 198, 203, 212, 225–227, 231, 239
control 5, 6, 13, 14, 24, 36, 46, 56, 59–61, 63, 86, 87, 89, 95, 117, 134, 165, 182, 187, 193, 198, 203, 205, 207, 209, 213, 214, 216, 219, 220, 223, 229, 232–239, 247, 249, 250, 256
Corn, Joe 15
cost 11, 36, 75, 76, 79, 81, 103, 110, 116, 119, 128, 166, 167, 178, 179, 183, 185, 206, 214
Croyden Airport 8
culture 10, 11, 13, 15, 24, 25, 31, 48, 74–76, 79, 97, 186, 191, 200, 201, 203, 217, 225, 230, 235, 250
customs 5, 13, 53, 56, 63, 83, 84, 89, 175, 176, 181, 195, 200, 210, 221, 223, 232, 233, 237

D

Dallas-Fort Worth Airport 192
dedication 47, 60, 65, 118
Delano & Aldrich 41, 43, 44, 49, 54, 55
Denver Airport 179, 214
design closure 2
designer 3, 8, 23, 42, 44, 53, 75, 105, 114, 115, 127, 179, 194, 245
Detroit Airport 184, 185
discourse 4, 10, 14, 60, 61, 71, 73, 75, 79, 128, 204, 206, 212, 233, 250
divorce 18
dock 11, 54, 74, 96
Douglas, Deborah 33, 129, 210
drop-off 55, 61, 89, 100, 106, 126, 128, 144
Dubai Airport 184, 200
Dublin Airport 8, 104, 183, 200
Dulles Airport 9, 11, 17, 74, 102, 110, 119, 120, 125, 135, 163, 177, 179, 189, 229, 254

E

East Boston Airport 29
edges 9, 156, 221, 234
Edinburgh Airport 104
engine 75, 94, 113
engineer 24, 34, 41, 51, 64, 77, 80, 98, 108, 115, 130, 150, 151, 166, 183, 189, 193
expert 4, 7, 16, 62, 72, 73, 105, 144, 145, 151, 154, 194, 209, 217, 220, 227, 235, 254

F

failure 76, 80, 160, 171, 177, 187, 254
Federal Aviation Authority (FAA) 11, 115, 193
fence 5, 28, 35, 94, 191
filter 5, 13, 14, 113, 203, 205, 210, 220, 234, 240, 250
finger 17, 82, 89, 90, 96, 101, 104, 105, 107, 109, 110, 112, 113, 116, 146, 148, 159, 162, 190, 246
flow 4, 9, 13, 28, 37, 53, 54, 56, 83, 85, 88, 94, 100, 137, 152,

180, 189, 203, 218, 222, 226, 227, 230, 232, 237, 239, 240, 250
Frankfurt Airport 212
Froesch, C. 75, 78, 79, 104
frontier 3, 6, 10, 13, 14, 17, 66, 89, 129, 156, 165, 175, 190, 192, 195, 196, 205, 206, 218–220, 226, 229, 234, 246, 254
functions 6, 9, 34, 40, 52, 53, 61, 78, 79, 131, 132, 147, 153

G

Gallison, Peter 15
gangplank 96, 97
Griesemer, James 4, 254
group 42, 87, 94, 115, 125, 127, 141, 146, 162, 164, 168, 169, 194, 198, 222

H

hangar 8
Harrison, Wallace K. 92, 187
Havana Airport 200
Heathrow Airport 101, 126, 238
Heterogeneous technical object 17, 96, 254
hippodrome 28
hold 4, 120, 178, 179, 182, 188, 193, 196, 200, 214, 226, 234
Hong Kong Airport 238
Hughes, Thomas 33, 75, 95, 120

I

Idlewild Airport 13, 65, 176, 185
immigration 83, 84, 200, 210, 224, 233, 237
imprinted 17, 47, 48, 63, 108, 126, 200, 254
Incheon Airport 238
information 7, 16, 49, 87, 111, 135, 159, 200, 212, 213, 216, 220, 221, 234, 237, 240
innovation 3, 11–13, 15, 22, 23, 27, 30, 34, 35, 49, 74, 107, 118, 127, 128, 136, 140, 165, 168, 171, 175, 177, 180, 182, 183, 185, 187, 188, 191, 196, 250, 255
international 3, 4, 6, 7, 9, 12–14, 72, 76, 84–86, 88, 89, 92, 96, 99, 101, 106, 108, 110, 113, 118, 119, 125–129, 132–138, 140, 141, 143–145, 147–149, 157, 160, 161, 163, 169, 176, 178–181, 183, 184, 188, 192–198, 200, 201, 204, 206, 207, 210, 211, 213–215, 218, 221–223, 225, 234, 235, 237, 238, 248
International Civil Aviation Organization (ICAO) 85, 86, 116, 213

J

Jeddah Airport 179
jet 3, 11, 73–75, 92, 94, 96, 100, 101, 107, 110, 112, 113, 117–120, 145, 164, 184, 248
jet-age 93, 108, 111
jetway 11, 74, 96
jurisdiction 86, 200

K
Kansai Airport 184
Kline, Ronald 31
Kuala Lumpur Airport 200

L
LaGuardia, Fiorello 37, 44, 60, 64
LaGuardia Airport 9, 21, 22, 47, 50, 56, 63–65, 85, 91, 126, 177, 187, 247, 254
landing 21, 27, 32, 33, 35, 50, 64, 97, 107, 113, 119, 137, 154, 214
landside 3–13, 16–18, 22, 27–29, 31, 39, 41, 47, 48, 50, 54, 60, 61, 63, 65, 66, 72–75, 77, 80–82, 84–92, 94–97, 99, 102, 104, 106, 108, 112–114, 116–121, 125, 127–129, 131, 132, 134–139, 141, 143, 144, 146–157, 160, 161, 165–171, 175, 176, 178–182, 184–188, 190–198, 200, 201, 203–207, 212, 213, 215–218, 220, 221, 223, 224, 226–239, 245–251, 253–255
large technological systems 16, 17, 33, 72, 217, 247, 254
Latour, Bruno 220
Leigh Fisher and Associates (LF&A) 12, 125, 127, 148, 166
Leipzig Airport 45
limbo 114, 164, 195, 196, 200, 211, 212, 217, 224, 225, 228, 238, 255
limits 5, 14, 18, 154, 182, 205, 217, 218, 229, 232, 236
linearity 17, 118, 183, 184, 246
liquid 13, 203, 219, 239, 250
Liverpool Airport 8, 45, 183
loading 8, 47, 50, 53, 54, 61, 81, 82, 95, 96, 98, 101, 105, 110, 147, 156, 162, 164
London Airport 107
London Oxford Airport 179
Long Beach Airport 104
Los Angeles Airport 105, 136, 179, 192

M
machine 24, 25, 27, 30, 36, 45, 47, 62, 64, 85, 94, 96, 107, 117, 148, 157, 177, 191, 209, 215, 217, 235, 254
Madrid Airport 184
Mehran, Nasseri 224
metis 105
metropolitan 42, 99, 116, 162, 196, 214
Mexico City Airport 200
Miami Airport 42, 43, 126, 165
Midway Airport 183
militarization 222
Mobile Lounge 11, 72, 74, 100, 104, 105, 107–110, 112–115, 118–121, 125, 127, 163, 164, 170, 177, 178, 248, 255
momentum 72, 73, 75, 84, 95, 97, 120, 248, 254
Monterrey Airport 8, 194, 195
Montreal Airport 104, 179

N

negotiation 4, 7, 8, 10, 16, 63, 81, 84, 87, 102, 108, 114, 117, 130, 132, 165–167, 170, 182, 194, 195, 212, 248
network 2, 36, 56, 61, 71, 76–78, 182, 207, 210–212, 254
Newark Airport 10, 37
New York Municipal 21, 40, 41, 49, 53, 55, 56, 65
nonplace 114, 225
non-sterile 215, 234
North Beach Airport 64
Nye, David 17, 24, 63, 248

O

O'Hare Airport 96, 143
Orly Airport 183, 188, 189, 208, 209, 238
Osaka Airport 184

P

Paris Charles de Gaulle Airport 12, 176, 185, 188, 225
Paris LeBourguet Airport 21, 208
parking 9, 28, 41, 50, 54, 55, 92, 103, 104, 120, 134, 138, 143, 144, 147, 151, 152, 180, 183, 184, 187–189, 230, 238
passenger 11, 32, 35, 37, 47, 51, 55, 61, 73, 74, 81–84, 86, 88, 90, 91, 93–95, 98, 100–105, 108, 110–114, 118, 126, 131, 135, 137, 140, 144, 146–149, 153, 157, 159, 161–164, 166, 170, 178, 179, 185, 188, 191, 195, 198, 210, 213–215, 223, 225–227, 230, 234, 238, 239, 247, 250
permeability 13, 203
Philadelphia Airport 104, 167, 179
physical 4, 5, 14, 18, 55, 57, 78, 83, 87, 89, 126, 144, 145, 147, 155, 156, 162, 182, 192, 197, 205, 207, 212, 213, 217–221, 228, 229, 231, 232, 234–237, 240, 246
pier 17, 22, 54, 82, 90, 92, 95, 96, 104, 105, 118, 146, 159, 190, 246
pilots 25, 27, 32, 66, 94, 102, 109, 112, 116, 119
Pinch, Trevor 17, 31
placelessness 14, 224–226, 256
plan 6, 8, 42, 53, 75, 76, 99, 107, 108, 110, 115, 140, 144, 146, 148, 159, 169, 170, 183, 187, 189, 192, 196, 197, 206, 248
planner 12, 51, 52, 78, 84, 91, 92, 98, 99, 106, 131–133, 144, 150, 152, 153, 179, 194
platform 8, 27, 45, 47, 50, 51, 53–55, 113, 117, 182, 189, 195
porosity 13, 203
problem solving 125, 126, 131, 169, 170, 248
processes 7, 15, 16, 34, 61, 62, 127, 165–167, 188, 196, 222, 248, 255

program 2, 26, 34, 40, 47, 50, 60, 76, 80, 81, 93, 104, 105, 115, 132, 133, 143, 146, 178, 180, 200
progress 11, 25, 56, 60, 62, 74–76, 79, 94, 100, 104, 105, 108, 111, 117, 118, 141, 145, 157, 162, 164, 248
Prokosch, W. 75, 78, 79, 104
Pudong Airport 200

R

rail 1, 2, 32, 39, 41, 45–48, 52, 54, 59, 61, 71, 80, 91, 134, 182, 210, 233
ramp 55, 96, 101, 104, 107, 118
reinvention 9, 10, 71–73, 99, 102, 125, 177, 180, 188, 191, 247–249, 255
retail 4, 14, 53, 192, 196, 198, 205, 226–228, 231, 238, 250
Reykjavik Airport 200
rhetorical closure 10, 16, 17, 157, 171, 249, 255
Rio de Janeiro Airport 104
Roche, Kevin 103
Rogers, Everett 177, 185
romantic 10, 18, 35, 57, 66, 95, 117, 141
Rome Airport 58, 106, 141, 193, 216
rooms 32, 60, 82, 83, 88, 91, 149, 178, 179, 182, 188, 194–196, 198, 200, 226, 228
Rosenberg, Nathan 15, 80
runway 28, 31, 35, 82, 107, 183, 196

S

Saarinen, Eero 11, 72, 74, 104, 107, 254
Safdie, Moshe 4, 194
safety 3, 28, 33, 34, 65, 119, 215, 233, 238
Saint Louis' Lambert Terminal 94, 178
San Francisco Airport 96, 134, 141, 148, 156, 166
Sao Paulo Airport 106
satellite 12, 17, 101, 104, 110, 137, 179, 189, 190, 246
Schiphol Airport 45, 184, 238
SCOT 17, 151, 157, 247, 255
Scott, James 87
sea 24, 32, 44, 52
seaport 49, 55, 102
Seattle Airport 104
security 14, 66, 89, 120, 136, 144, 155, 156, 186, 190, 192, 193, 195–198, 200, 204, 206–210, 212–215, 217–222, 225–227, 229, 232–234, 236–239, 250, 253
segregation 4, 5, 45, 61, 84, 87, 89, 184, 194, 222, 225
separation 4, 6, 13, 53, 55, 56, 63, 83, 89, 104, 116, 131, 132, 166, 194, 203, 222, 228, 248
Shanghai Airport 200
Shannon Airport 200
Singapore Airport 200, 211, 238
Sismondo, Sergio 15
Skidmore, Owings and Merrill (SOM) 92, 179
Skywalk 10, 23, 54–60, 62–64, 66, 105, 126, 177, 178, 247

social 2, 3, 14, 17, 18, 23, 30, 31, 42, 44, 48, 51, 58, 63, 74, 84, 94, 97, 99, 115, 118, 120, 126, 127, 151, 157, 165, 168, 175, 185, 197, 205, 208, 210, 211, 218, 219, 222–224, 226, 233, 236, 240, 247, 248, 253–255
soft transition 31, 39, 45, 48, 56, 63, 247
sorting 49, 213
sovereignty 225
space 9, 13, 23, 31, 36, 46, 55, 61, 79–82, 84, 86, 89, 90, 103, 104, 107, 110, 140, 147, 162, 180, 193, 195, 196, 208, 209, 215, 216, 219, 220, 225–228, 231, 232, 246, 250, 255, 256
specialist 50, 52, 81, 91, 116, 212
spectators 23–26, 28, 29, 48, 57, 60, 63, 83, 84, 86
speed 3, 15, 34, 37, 130, 157, 161, 162, 178, 180, 209, 222
Speke Airport 45, 183
standard 3, 12, 17, 48, 58, 73, 77, 86, 88, 105, 120, 126, 131, 143, 152, 169–171, 175, 182, 201, 215, 223, 234, 246–249, 253, 254
Stansted Airport 179
Star, Susan L. 3, 254
station 1, 32, 46, 47, 54, 59, 61, 80, 148, 182, 210, 233, 235
sterile 215, 234
St. Louis Airport 106, 115
suburban 46, 93
surveillance 14, 18, 182, 212, 215, 220, 221, 225, 228, 232, 235, 238, 256
Sydney Airport 104, 224
symmetry 16
system builder 17, 103, 114, 132, 165, 249, 254
systems 3, 7–10, 13, 14, 31, 33, 34, 46, 48, 49, 51, 52, 55, 56, 61, 63, 73, 75, 77, 95, 99, 101, 104, 106, 110, 114, 143, 147, 152, 162, 163, 176, 177, 183, 184, 191, 208, 210, 212, 217, 219, 220, 228, 229, 234, 238, 239, 250, 253–255

T

Tampa Airport 137, 140, 143, 154
tangible objects 4
technological frame 17, 108, 127, 128, 165, 168, 248, 255
technological sublime 17, 24, 63, 254
Tel Aviv Airport 198
Tempelhof Airport 8
terminal 3, 7, 12, 14, 18, 21, 32, 33, 35, 40, 41, 43, 46, 47, 50, 53, 54, 57–60, 63, 64, 72, 77, 81, 82, 85–88, 90, 92, 94–96, 98, 101, 102, 106–108, 110–114, 116–120, 128, 130–143, 145–153, 156–165, 171, 176–179, 182–193, 195, 197, 200, 201, 205, 208–210, 212, 214–216, 224, 227, 228, 231, 239, 246
Tokyo Airport 143, 179
Toluca Airport 194, 196, 197
Toronto Airport 4, 194, 195

traffic 35, 46, 55, 56, 75, 77, 81, 82, 92–94, 97, 100, 101, 104, 108, 119, 126, 130, 134, 136, 137, 159, 161, 170, 181, 214, 221, 225, 230
trajectories 80, 82
transnational 224, 238
transparency 13, 175, 239
transportation 2, 12, 36, 43, 45, 46, 54, 59, 60, 62, 76–78, 88, 93, 104, 110, 116, 128, 137, 143, 147–152, 161, 162, 164, 167, 180, 183, 187, 197, 218, 222, 234, 245
travel 2, 32, 37, 44, 46, 66, 74, 78, 84, 86, 87, 92, 93, 100, 105, 115, 118–120, 127, 143, 162, 180, 181, 198, 215, 223, 227, 230, 231, 234, 240, 254
Turkle, Sherry 217
TWA Terminal 11, 74, 94, 98, 178, 209
typology 17, 73, 82, 176, 246

U

uncertainty 15, 78, 171, 185, 195, 228, 249
urban 8, 11, 26, 27, 30, 31, 74, 78, 79, 93, 230, 238
users 3, 10, 17, 63, 74, 84, 87, 93, 95, 100, 103, 110, 115–117, 128, 135, 138, 139, 150, 151, 154, 165, 170, 181, 190, 192, 196, 203, 206, 208, 211, 213, 214, 216, 217, 220, 221, 226, 228–232, 236, 237, 246, 248, 250, 251, 255

V

valves 13, 171, 194, 203, 232
Vancouver Airport 200, 201
Vienna Airport 212
viewport 216–218, 250
virtual 14, 205, 207, 217, 220, 229, 234–236, 240, 246, 255, 256
visitors 2, 23, 28, 41, 47, 51, 56–58, 60, 81–84, 88, 95, 139, 141, 226, 229, 237

W

Washington Airport 11, 72
Winner, Langdon 60
Work Projects Administration (WPA) 52, 77

Z

zone 148, 150, 171, 195, 196, 211, 217, 218, 223, 224, 226, 228, 230
Zurich Airport 104, 106, 178

The manufacturer's authorised representative in the EU is Springer Nature Customer Service Centre GmbH, Europaplatz 3, 69115 Heidelberg, Germany. If you have any concerns regarding our products, please contact ProductSafety@springernature.com

Printed and bound by CPI Group (UK) Ltd, Croydon, CR0 4YY
30/09/2026
02225451-0012